Elephant Don

The Politics of a Pachyderm Posse

By Caitlin O'Connell

These are UNCORRECTED page proofs.
Not for sale or distribution.
No part of these page proofs may be
reproduced in any form or quoted
without the written permission of
the University of Chicago Press.

ISBN-13: 978-0-226-10611-3 (Cloth)

$26.00   £18.00

256 p., 44 halftones 6 x 9

Publication date: April 2015

For additional information, contact:

Carrie Olivia Adams
Assistant Promotions Director
The University of Chicago Press
1427 East 60th Street
Chicago, IL 60637
(773) 702-4216
fax (773) 702-9756
cadams@press.uchicago.edu

Elephant Don

▾▲▾▲▾▲▾▲▾

Elephant Don

The Politics of a Pachyderm Posse

▼▲▼▲▼▲▼▲▼

CAITLIN O'CONNELL

The University of Chicago Press
Chicago and London

{~¿~**Author** bio to come from marketing.}

The University of Chicago Press, Chicago 60637
The University of Chicago Press, Ltd., London

24 23 22 21 20 19 18 17 16 15   1 2 3 4 5

ISBN-13: 978-0-226-10611-3 (cloth)
ISBN-13: 978-0-226-10625-0 (e-book)
DOI: 10.7208/chicago/9780226106250.001.0001

CIP data to come

⊗ This paper meets the requirements of ANSI/NISO
Z39.48-1992 (Permanence of Paper).

I would like to dedicate this book to Greg, the Elephant Don—aptly represented by the Lozi proverb on the next page, since his male associates can, while he is "holding court" with them, either protect or harm him. May Greg be out there somewhere remembering the good old days among his loyal companions, who still appear to be looking for him. Perhaps one day he will return to reward their vigilance. But if he doesn't, I hope that his character will remain an example to his associates—and to us—of what a fine mentor one can be to the next generation, given a combination of patience, affability, wisdom, and a good firm trunk.

When the King is in the court, he is an elephant in thorns.

Lozi Proverb, Zambia

CONTENTS

——— ▼▲▼▲▼▲▼ ———

Kissing of the Ring

▼▲▼▲▼▲▼▲▼

Sitting in our research tower at the water hole, I sipped my tea and enjoyed the late morning view. A couple of lappet-faced vultures climbed a nearby thermal in the white sky. A small dust devil of sand, dry brush, and elephant dung whirled around the pan, scattering a flock of guinea fowl in its path. It appeared to be just another day for all the denizens of Mushara water hole—except the elephants. For them, a storm of epic proportions was brewing.

It was the beginning of the 2005 season at my field site in Etosha National Park, Namibia—just after the rainy period, when more elephants would be coming to Mushara in search of water—and I was focused on sorting out the dynamics of the resident male elephant society. I was determined to see if male elephants operated under different rules here than in other environments and how this male society compared to other male societies in general. Among the many questions I wanted to answer was how ranking was determined and maintained and for how long the dominant bull could hold his position at the top of the hierarchy.

While observing eight members of the local boys' club arrive for a drink, I immediately noticed that something was amiss—these bulls weren't quite up to their usual friendly antics. There was an undeniable edge to the mood of the group.

The two youngest bulls, Osh and Vincent Van Gogh, kept shifting their weight back and forth from shoulder to shoulder, seemingly looking for reassurance from their mid- and high-ranking elders. Occasionally, one or the other held its trunk tentatively outward—as if to gain comfort from a ritualized trunk-to-mouth greeting.

The elders completely ignored these gestures, offering none of the usual reassurances such as a trunk-to-mouth in return or an ear over a youngster's head or rear. Instead, everyone kept an eye on Greg, the most dominant member of the group. And for whatever reason, Greg was in a foul temper. He moved as if ants were crawling under his skin.

Like many other animals, elephants form a strict hierarchy to reduce conflict over scarce resources, such as water, food, and mates. In this desert environment, it made sense that these bulls would form a pecking order to reduce the amount of conflict surrounding access to water, particularly the cleanest water.

At Mushara water hole, the best water comes up from the outflow of an artesian well, which is funneled into a cement trough at a particular point. As clean water is more palatable to the elephant and as access to the best drinking spot is driven by dominance, scoring of rank in most cases is made fairly simple—based on the number of times one bull wins a contest with another by usurping his position at the water hole, by forcing him to move to a less desirable position in terms of water quality, or by changing trajectory away from better-quality water through physical contact or visual cues.

Cynthia Moss and her colleagues had been figured out a great deal about dominance in matriarchal family groups by. Their long-term studies in Amboseli National Park showed that the top position in the family was passed on to the next oldest and wisest female, rather than to the offspring of the most dominant individual. Females formed extended social networks, with the strongest bonds being found within the family group. Then the network branched out into bond groups, and beyond that into associated groups called clans. Branches of these networks were fluid in nature, with some group members coming together and others spreading out to join more distantly related groups in what had been termed a fission-fusion society.

Not as much research had been done on the social lives males, outside the work by Poole and her colleagues in the context of musth and one-on-one contests. I wanted to understand how male relationships were structured after leaving their maternal family groups as teens, when much of their adult lives was spent away from their female family. In my previous field seasons at Mushara, I'd noticed that male elephants formed much larger and more consistent groups than had been reported elsewhere and that, in dry years, lone bulls were not as common here than were recorded in other research sites.

Bulls of all ages were remarkably affiliative—or friendly—within associated groups at Mushara. This was particularly true of adolescent bulls, which were always touching each other and often maintained body contact for long periods. And it was common to see a gathering of elephant bulls arrive together in one long dusty line of gray boulders that rose from the tree line and slowly morphed into elephants. Most often, they'd leave in a similar manner—just as the family groups of females did.

The dominant bull, Greg, most often at the head of the line, is distinguishable by the two square-shaped notches out of the lower portion of his left ear. But there is something deeper that differentiates him, something that exhibits his character and makes him visible from a long way off. This guy has the confidence of royalty—the way he holds his head, his casual swagger: he is made of kingly stuff. And it is clear that the others acknowledge his royal rank as his position is reinforced every time he struts up to the water hole to drink.

Without fail, when Greg approaches, the other bulls slowly back away, allowing him access to the best, purest water at the head of the trough—the score having been settled at some earlier period, as this deference is triggered without challenge or contest almost every time. The head of the trough is equivalent to the end of the table and is clearly reserved for the top-ranking elephant—the one I can't help but refer to as the *don* since his subordinates line up to place their trunks in his mouth as if kissing a Mafioso don's ring.

As I watched Greg settle in to drink, each bull approached in turn with trunk outstretched, quivering in trepidation, dipping the tip into Greg's mouth. It was clearly an act of great intent, a symbolic gesture of respect for the highest-ranking male. After performing the ritual, the lesser bulls seemed to relax their shoulder as they shifted to a lower-ranking position within the elephantine equivalent of a social club. Each bull paid their respects and then retreated. It was an event that never failed to impress me—one of those reminders in life that maybe humans are not as special in our social complexity as we sometimes like to think—or at least that other animals may be equally complex. This male culture was steeped in ritual.

Greg takes on Kevin. Both bulls face each other squarely, with ears held out. Greg's ear cutout pattern in the left ear make him very recognizable

But today, no amount of ritual would placate the don. Greg was clearly agitated. He was shifting his weight from one front foot to the other in jerky movements and spinning his head around to watch his back, as if someone had tapped him on the shoulder in a bar, trying to pick a fight.

The midranking bulls were in a state of upheaval in the presence of their pissed-off don. Each seemed to be demonstrating good relations with key higher-ranking individuals through body contact. Osh leaned against Torn Trunk on his one side, and Dave leaned in from the other, placing his trunk in Torn Trunk's mouth. The most sought-after connection was with Greg himself, of course, who normally allowed lower-ranking individuals like Tim to drink at the dominant position with him.

Greg, however, was in no mood for the brotherly "back slapping" that ordinarily took place. Tim, as a result, didn't display the confidence that he generally had in Greg's presence. He stood cowering at the lowest-ranking position at the trough, sucking his trunk, as if uncertain of how to negotiate his place in the hierarchy without the protection of the don.

Finally, the explanation for all of the chaos strode in on four legs. It was Kevin, the third-ranking bull. His wide-splayed tusks, perfect ears, and bald tail made him easy to identify. And he exhibited the telltale sign of musth, as urine was dribbling from his penis sheath. With shoulders high and head up, he was ready to take Greg on.

A bull entering the hormonal state of musth was supposed to experience a kind of "Popeye effect" that trumped established dominance patterns—even the alpha male wouldn't risk challenging a bull elephant with the testosterone equivalent of a can of spinach on board. In fact, there are reports of musth bulls having on the order of twenty times the normal amount of testosterone circulating in their blood. That's a lot of spinach.

Musth manifests itself in a suite of exaggerated aggressive displays, including curling the trunk across the brow with ears waving—presumably to facilitate the wafting of a musthy secretion from glands in the temporal region—all the while dribbling urine. The message is the elephant equivalent of "don't even think about messing with me 'cause I'm so crazy-mad that I'll tear your frickin' head off"—a kind of Dennis Hopper approach to negotiating space.

Musth—a Hindi word derived from the Persian and Urdu word "mast," meaning intoxicated—was first noted in the Asian elephant. In Sufi philosophy, a mast (pronounced "must") was someone so overcome with love for God that in their ecstasy they appeared to be disoriented. The testosterone-heightened state of musth is similar to the phenomenon of rutting in antelopes, in which all adult males compete for access to females under the influence of a similar surge of testosterone that lasts throughout a discrete season. During the rutting season, roaring red deer and bugling elk, for example, aggressively fight off other males in rut and do their best to corral and defend their harems in order to mate with as many does as possible.

The curious thing about elephants, however, is that only a few bulls go into musth at any one time throughout the year. This means that there is no discrete season when all bulls are simultaneously vying for mates. The prevailing theory is that this staggering of bulls entering musth allows lower-ranking males to gain a temporary competitive advantage over others of higher rank by becoming so acutely agitated that dominant bulls wouldn't want to contend with such a challenge, even in the presence of an estrus female who is ready to mate. This serves to spread the wealth in terms of gene pool variation, in that the dominant bull won't then be the only father in the region.

Given what was known about musth, I fully expected Greg to get the daylights beaten out of him. Everything I had read suggested that when a top-ranking bull went up against a rival that was in musth, the rival would win.

What makes the stakes especially high for elephant bulls is the fact that estrus is so infrequent among elephant cows. Since gestation lasts twenty-two months, and calves are only weaned after two years, estrus cycles are spaced at least four and as many as six years apart. Because of this unusually long interval, relatively few female elephants are ovulating in any one season. The competition for access to cows is stiffer than in most other mammalian societies, where almost all mature females would be available to mate in any one year. To complicate matters, sexually mature bulls don't live within matriarchal family groups and elephants range widely in search of water and forage, so *finding* an estrus female is that much more of a challenge for a bull.

Long-term studies in Amboseli indicated that the more dominant bulls still had an advantage, in that they tended to come into musth when more females were likely to be in estrus. Moreover, these bulls were able to maintain their musth period for a longer time than the younger, less dominant bulls. Although estrus was not supposed to be synchronous in females, more females tended to come into estrus at the end of the wet season, with babies appearing toward the middle of the wet season, twenty-two months later. So being in musth in this prime period was clearly an advantage.

Even if Greg enjoyed the luxury of being in musth during the peak of estrus females, this was not his season. According to the prevailing theory, and in this situation, Greg would back down to Kevin.

As Kevin sauntered up to the water hole, the rest of the bulls backed away like a crowd avoiding a street fight. Except for Greg. Not only did Greg not back down, he marched clear around the pan with his head held to its fullest height, back arched, heading straight for Kevin. Even more surprising, when Kevin saw Greg approach him with this aggressive posture, he immediately started to back up.

Backing up is rarely a graceful procedure for any animal, and I had certainly never seen an elephant back up so sure-footedly. But there was Kevin, keeping his same even and wide gait, only in the reverse direction— like a four-legged Michael Jackson doing the moon walk. He walked backward with such purpose and poise that I couldn't help but feel that I was watching a videotape playing in reverse—that Nordic-track style gait, fluidly moving in the opposite direction, first the legs on the one side, then on the other, always hind foot first.

Greg stepped up his game a notch as Kevin readied himself in his now fifty-yard retreat, squaring off to face his assailant head on. Greg puffed up like a bruiser and picked up his pace, kicking dust in all directions. Just before reaching Kevin, Greg lifted his head even higher and made a full frontal attack, lunging at the offending beast, thrusting his head forward, ready to come to blows.

In another instant, two mighty heads collided in a dusty clash. Tusks met in an explosive crack, with trunks tucked under bellies to stay clear of the collisions. Greg's ears were pinched in the horizontal position—an extremely aggressive posture. And using the full weight of his body, he raised his head again and slammed at Kevin with his broken tusks. Dust flew as the musth bull now went in full backward retreat.

Amazingly, this third-ranking bull, doped up with the elephant equivalent of PCP, was getting his hide kicked. That wasn't supposed to happen.

At first, it looked as if it would be over without much of a fight. Then, Kevin made his move and went from retreat to confrontation and approached Greg, holding his head high. With heads now aligned and only inches apart, the two bulls locked eyes and squared up again, muscles tense. It was like watching two cowboys face off in a western.

There were a lot of false starts, mock charges from inches away, and all manner of insults cast through stiff trunks and arched backs. For a while, these two seemed equally matched, and the fight turned into a stalemate.

But after holding his own for half an hour, Kevin's strength, or confidence, visibly waned—a change that did not go unnoticed by Greg, who took full advantage of the situation. Aggressively dragging his trunk on the ground as he stomped forward, Greg continued to threaten Kevin with body language until finally the lesser bull was able to put a man-made structure between them, a cement bunker that we used for ground-level observations. Now, the two cowboys seemed more like sumo wrestlers, feet stamping in a sideways dance, thrusting their jaws out at each other in threat.

The two bulls faced each other over the cement bunker and postured back and forth, Greg tossing his trunk across the three-meter divide in frustration, until he was at last able to break the standoff, getting Kevin out in the open again. Without the obstacle between them, Kevin couldn't turn sideways to retreat, as that would have left his body vulnerable to Greg's formidable tusks. He eventually walked backward until he was driven out of the clearing, defeated.

In less than an hour, Greg, the dominant bull displaced a high-ranking bull in musth. Kevin's hormonal state not only failed to intimidate Greg but in fact just the opposite occurred: Kevin's state appeared to fuel Greg into a fit of violence. Greg would not tolerate a usurpation of his power.

Did Greg have a superpower that somehow trumped musth? Or could he only achieve this feat as the most dominant individual *within* his bonded band of brothers? Perhaps paying respects to the don was a little more expensive than a kiss of the ring.

Journey to Mushara

▼▲▼▲▼▲▼▲▼

Mushara water hole is an oasis where the pattern of animal movements demarks the passage of time almost as reliably as the cycles of the sun and moon. By day, you can enjoy a breathtaking diversity of wildlife with giraffe necks cluttering the horizon, oryx and springbok males sparring in the clearing, and the zigzag pattern of zebra reflections shimmering in the center of the pan, while the crowned and blacksmith lapwings patrol the edges.

By night, you can witness grumpy male rhinos in the heat of a territorial battle and lions stalking an elephant family group as they panic to protect their young. All this under an obsidian sky that appears to be a gateway to the universe—the Milky Way so bright, it reflects in the pan and seems to cast starlight on the white sand glistening below.

If you want to spend the days on driving safaris and return to a hot shower, flush toilets, and a steak dinner served on china, Mushara might not be for you. I, however, find the inevitable grit of sand between my teeth at this remote wildlife mecca to be worth every grain.

The preparations in the capital city, Windhoek, prior to our arrival at Mushara every year became more and more routine. After a trip to the hardware store for nuts and bolts, buying an inordinate amount of canned and dried food at the Trade Center, and procuring several hundred pounds of potatoes, butternut squash, gem squash, carrots, onions, English cucumber, and cabbage at the Fruit & Veg Market, we were good to go.

Despite not being able to get everything on our list that we had wanted, we always made peace with what we were able to obtain. Then, my co-expedition leader and husband, Tim, and I would collect our team at the airport and, schedule permitting, treat them to a safari-style dinner around an open fire at the local Joe's Beer House before making our way north early the next morning.

We usually spent the first night of our safari at Okaukuejo, the main tourist rest camp in Etosha that also supports a small research camp. A five-hour drive north from Windhoek, Okaukuejo is a good stopover, since our study site was too far away to allow us to drive and set up camp within a single day. Stopping at the research camp at Okaukuejo also gave us a chance to catch up with the local research staff and coordinate our schedule with them. It had the added advantage of a floodlit water hole that was very active at night.

The ability to sit for a moment and watch elephants at sunset against a pink backdrop, white dust hovering above the calcrete rocky ground always made the frustrations of packing and organizing melt away. Chalky elephant bulls stood at the water hole, and inevitably a dominant bull would shove another away from the source of the spring. The subordinate would bellow in objection, but then settle in to drink again a safe distance away. If we were lucky, a family group would come in for a drink as well, making the sunset all the more special.

The view from the Okaukuejo water hole never got old; it just became increasingly like a home to which I felt compelled to return. I had spent my thirtieth birthday here, and then my fortieth, and next year, my fiftieth. But sometimes, in the difficult months of planning that led up to the field season, it was easy to forget the magic of it all. It took these moments at the water hole to bring back the excitement and anticipation of the coming season.

At dusk, we'd reluctantly leave the water hole to make a quick dinner of something easy like vegetable curry and couscous at the research camp. After dinner, despite our fatigue, we'd often return as the team would be eager to see their first black rhino, and so we'd drive back to the floodlit water hole for a quick nightcap of wildlife.

When we arrived at the water hole at the beginning of the 2005 season, we were greeted with the sight of two elephant bulls and seven black rhinos. After a few gasps at so many of this rare species in one sighting, everyone settled in to take in the view, sitting on one of the few benches unoccupied by the other hushed park visitors.

As I watched an old gnarled bull elephant drink from within his black reflection, I zipped up my down jacket and wondered what I'd be thinking, standing in this same place two months from then. Because it had been a very low rainfall year and there would be few other places for animals to drink in the region, I was expecting a lot of action at Mushara this season.

While admiring the old bull, his flatulence and all, I found myself revisiting my first night at Okaukuejo water hole back in 1992 around this same time in early June. I was an entomologist back then, and Tim and I had just come up from the Namib Desert where I was awestruck by the diversity of ground beetles, which resembled dune buggies in that their very long legs distanced them from the scorching desert sand.

We had spent the month prior to that in the Karoo Desert and Cape fynbos in South Africa, marveling over dinosaur footprints frozen in time and geological formations I had only read about in textbooks. Sun birds dipped in and out of the long trumpet-shaped *Erica* flowers with their metallic green and blue wings shining in the sun. Protea flowers looked like flamingo feathers, and bee flies came in shapes and sizes I had never seen. The natural history and animal diversity of Africa stole any ambition I'd had to study anywhere else.

Back then, Tim and I had no particular plan for how to spend our nine-month vacation between graduate degrees, except for the idea that we'd drive our '73 Beetle all the way from Cape Town to Kenya. But after that first night at Okaukuejo, my thirst for Africa took a different form.

There was something about seeing these great ghostly white giants coming and going silently throughout the night that made me want to understand Africa on a much more personal level. I couldn't bear the thought of being a tourist in this amazingly intact land that was still so wild and full of life. Unexpectedly, the course of my personal and professional life and the shape of the next twenty-five years of my career were changed the very next day: within a week of volunteering at the Etosha Ecological Institute, we were offered a three-year contract to study elephants for the Namibian government. Out of the blue, my relationship with elephants began and has kept me coming back to this place—to the elephants of Etosha.

The next morning we woke early to the frigid June air and had coffee and rusks (Afrikaans equivalent of biscotti) on the go. We stopped to pack up the remaining equipment that we had stored off season at the research camp and gassed up the truck we had borrowed from the ministry.

Considering our slow start, we were still on a good schedule, pulling out of Okaukuejo at seven-thirty in the morning, only an hour after the park gates opened at sunrise. We had a three-hour drive ahead of us and then a long day of setting up camp. We made our way east across the park in a three-truck caravan.

Proclaimed as a game reserve in 1907 by German Governor Friedrich von Lindequist, Etosha was originally a little more than ninety thousand square kilometers, incorporating the massive extinct lake, Etosha Pan, which spans 4,590 kilometers. The size of the park fluctuated over the next century, and the current area covers just over twenty-two thousand square kilometers and hosts 114 mammal species, 340 bird species, 110 reptiles, 16 amphibians, and one species of fish.

The park is relatively flat except for rocky escarpments in the west that are closed to tourism. Much of the rest of the park ranges in habitat from vast grass plains and shrubs surrounding the pan, to deciduous mopane trees and mixed bushveld, to the tree sandveld of the northeast, where taller trees grow in the deep sand. The taller trees define leopard country.

We passed through the golden Halali Plains, the resident springbok dotting the open space from foreground to horizon. A disturbance far out to the left sent a large number of springbok "pronking" their way to the right of the landscape, heads down, backs arched as they popped straight up and down like popcorn in huge stiff-legged leaps, their fluffy tails and the hair on their backsides erect and gleaming white in the distance. I imagined cheetahs hiding out in the patches of silvery catophractes, or rattle pod bushes, their favorite hunting grounds, licking their chops at the thought of a yearling springbok for breakfast.

As we drove on, wildebeests, sitting in open grassland chewing their cud, came into view, and zebras clustered into their harems. Some formed shade for those who lay on their sides with stiff legs and bloated shining white bellies, looking as if they were in the first stages of rigor mortis.

Even if in poor physical condition, zebras always look fat due to the large amount of gas they generate in their gut through fermentation of the poor-quality food they have access to in order to extract as many nutrients as possible from it. Had we not inadvertently startled one of these prostrate creatures out of its sunny slumber in the middle of the road, I would have wondered whether they were indeed dead.

We finally reached the gleaming bleach-white calcrete of the pan's edge and stopped to take a look from a rest stop where tourists were allowed to get out, stretch their legs, and use the bathroom. Muddy ostrich tracks from the last wet season were baked into the parched and cracked clay, traversing the impossibly white and empty space for as far as the eye could see, the distance blurring into a mirage. The tracks looked every bit as if they were made by a two-toed dinosaur having awakened from a prehistoric slumber to take a stroll in this desolate sinking pit.

By late morning we arrived in Namutoni, site of an old German fort and a reminder of a tumultuous past during Europe's scramble for Africa. Namibia was the German colony of Southwest Africa before being turned over to South Africa as a protectorate after World War I. The brick fort resembled a white castle surrounded by makalani palms and seemed incongruous among the giraffe and scrub bush.

We were making good time. At Namutoni, we stopped to check in with the local rangers and, after receiving a warm greeting, turned north toward the Pleistocene-like Andoni Plains.

We first crossed Fisher's Pan, which in the wetter years still harbored flamingos at this time of year. But today we saw just the odd shelduck pair, a pygmy goose, and some red-billed teals bobbing in the shallows.

As we drove on, the open pan quickly turned to scrub and then to the deeper sandveld where the large purple papery-podded trees that the Owambo people call Mushara dominate, hence the name of the water hole at our field site. We started to see the first signs of elephants in the area—large old calcrete-powdered dung boluses lining the dusty clay road. The quantity increased as we approached the water hole called Tsumcor, a favorite stop for elephants. But when we turned down the track that led to the water hole we found only a lone male kudu, his horns making three long sweeping twists into the empty sky.

By lunchtime, we'd reached the firebreak road that led to Mushara and Kameeldoring. We turned right onto the deep sandy track and locked our wheels to engage four-wheel drive. The sand was so hot it felt soupy under the wheel, as we wallowed down the track until we reached a thick sandy T-junction, where we turned north toward Mushara.

As we got close to Mushara, we could see our observation tower in the distance, just visible above the tree line. It was leaning terribly askew. The elephants had been up to their usual tricks. The tower was getting fairly dilapidated, and I knew we'd have to face the necessity of building a new one in the near future. But for the present, given our considerable financial constraints and limited time, this was all we had to work with.

Nevertheless, my heart sank. Though we had our work cut out for us, at least we had anticipated as much in considering a worst-case scenario and brought everything we needed for the repairs. We would also have the expertise of the park's research technician, Johannes Kapner. Over the years of working in the park, he had his share of building and maintaining structures in remote places where there were lots of large animals. He knew just what to expect of an elephant's curiosity under these conditions and how to design a research camp accordingly.

I was happy to see that the dry-season traffic had already started, as was evident from the amount of elephant dung scattered throughout the clearing. I planned to retrieve dung for hormone analysis and genetic samples, but judging from the amount that had been scattered around the site, I was going to have some competition: guinea fowl find fresh elephant dung irresistible for dust bathing and for the copious amount of partially digested seeds within, and they had been busy ruffling up their feathered skirts with sweet excrement.

Off to the left, I spotted a pair of lions relaxing in the bush. At our approach, they lifted their heads and glared at us through large golden eyes. But they quickly laid their heads back down, eyes closed, uninterested in our presence.

We posted someone on the lookout while unloading, as it was easy to get focused on our camp-building tasks and forget the necessity for vigilance against lions. Fortunately, a lion by day is a much different animal than a lion at night, and the dozing male and female remained where they were, most probably more interested in their own consort than they were in us.

After unloading some of our toolboxes, we got to work straightening the tower, using a vehicle and winch and a long sturdy pole to prop up the leaning side. Then we straightened the campsite's four metal corner posts. Elephants loved to use these as scratching posts, and inevitably we had to pull them up straight at the beginning of every season.

The next task in getting the site up and running was securing the camp with boma cloth, a lightweight, opaque, tan-colored material that we used as a wind and visual screen. The two-meter high boma cloth hid our movements and activities throughout the season so that we would neither disturb them nor draw their attention to us. Ironically, we were using the boma cloth to enclose ourselves rather than the wildlife.

Bobtail's pride prepares for a morning hunt.

Giraffes, in particular, are perturbed by any unusual activity. Their keen eyesight enables them to see each other at great distances as well as to assess danger. Any movement out of the norm could cause them to wait at the edge of the clearing for hours before finding it safe enough to approach the water. And because of their long necks, giraffes have to position themselves in a very vulnerable posture when they drink, so they need to be certain that conditions are safe before doing so. Our use of the boma cloth made for a less stressful experience for them while we went about our research activities.

Over the course of the morning, we periodically checked on the honeymooning lion pair, but since they continued to be more focused on themselves than on us, we became more comfortable with their presence. Every once in a while, the female would roll onto her back, paws in the air, without a hint of interest in our half-built camp.

By late afternoon, we'd strung five strands of wire around the four heavy metal corner posts to attach our cloth and enclose the ten-by-ten-meter camp. We had also dug in some intermediate wooden posts for added strength. Then later, when the wind died down, we secured the cloth onto the wire all around the fence.

After digging a long-drop outhouse (i.e., one with a very deep hole), we assembled our kitchen tent, tables, mess kits, coolers, cast-iron pots and everything else that we used to cook our meals, including two hundred pounds of butternut squash—the best staple fresh food in a place with no refrigeration.

In previous years, I wasn't sure how the squash would fare with the resident elephant population, having witnessed elephants raiding campsites near Victoria Falls, Zimbabwe, in search of potatoes, squash, and citrus. It seemed that the elephants of Mushara had not yet engaged in such bad behavior, nonetheless I was still careful about fresh fruit, knowing that would be particularly tempting fare in this arid place.

Finally, we set up our roof tent platforms, two meters off the ground and mounted the tents into place. We each then set up our individual tents and unpacked a few essentials before it got too dark. It was a relief to have most of the hard work finished and to get down to making our own little cozy homes in which we would be living for the next two months.

It was just getting dark by the time we finished. After a quick dinner of rice and tinned vegetable curry, we sat up in the tower, enjoying the silvery light of a half moon over the landscape as the lion couple came over to inspect our digs. They enjoyed a little more honeymooning in front of camp—his orgasmic capitulations sounding every bit like regurgitations.

We gave up long before the lions did and went to bed, where we were treated to their repetitive noisy fornications every six or so minutes for much of the rest of the night. Since ovulation is induced by copulation, the lioness's intense interest in exhausting her partner with performance requests on the order of a thousand in number (with as many as three thousand interactions reportedly occurring during a lioness's estrus cycle) was not without evolutionary value.

The wee hours were also filled with the whooping calls of a new den of hyenas that had settled in the area. Although not a peaceful night's sleep, with the camp finally in order, it was the most restful sleep I had had in months.

After breakfast the next morning, we set about the construction of a second platform for the tower. Although I was primarily focused on studying social interactions among male elephants, I also planned to conduct ongoing experiments on elephant vocalizations and seismic communication, and a second story in the tower allowed our research assistants to collect behavioral data while remaining unaware of the timing and content of our playback experiments. Once the second floor was secured, we set up a pulley system to haul tables, chairs, and equipment up to both the observation and research platforms.

Next, we set up our solar-energy charging system: two eighty-watt solar panels and six car batteries—one pair of batteries for each floor of the tower and a third pair for the ground. We would rely on these rechargeable car batteries to charge video and camera batteries and computers and to power our tiny twelve-volt refrigerator, where we kept a few precious dairy products and cold drinks to have at sunset, as a "sundowner" as the locals liked to say. The batteries also ran our handmade elephant-dung dryer on the ground. The dryer worked like a desiccator used to dry fruit, equipped with separate mesh drying racks and numbered baskets to keep track of whose dung was whose during the drying process.

Over the next few weeks, we slowly built up our identification book of resident bulls, which already included fifty elephants that we could now recognize by various physical features as well as by approximate age—categorized as quarter, half, three-quarter, and full-sized bulls. Then we gave each bull a catalog number as well as a name that was indicative of either his physical features or overall character.

The bunker is twenty meters from the water hole and is buried, providing a military-like pillbox view at ground level. Sometimes the occupants of the bunker are more interesting to the wildlife than the other way around.

To the untrained eye, bull elephants look pretty much the same—they are big and gray, have a trunk, tusks, and a tail (and, of course, a very large penis, if it is extended). Some bulls are obviously bigger, some smaller, and some have larger tusks than others. But, on closer investigation, each bull had unique physical characteristics. There were invariably ear-notch patterns, tusk shapes, tail-hair patterns, and other features with pronounced or subtle differences among individuals that enabled us to tell them apart.

We used a laser altimeter to measure shoulder height, which in turn scales with tooth wearing and number of molars, the best estimate of age. Elephants have a total of six sets of teeth that erupt during their lifetimes, with only three molars present in each quadrant at any one time. Depending on the habitat and type of food available, when an elephant is in his fifties to sixties, his last set of molars erupt, then eventually wear down and drop out. At this point he is no longer able to chew food and before long dies a natural death. So, from this previous work on how elephants' teeth wear over time, other researchers developed a metric for correlating tooth wear with shoulder height. This data allowed us to generate a good estimate of a bull's age simply by measuring shoulder height.

When elephants left the area, we also measured hind-foot length from individual footprints to pinpoint age more accurately, since the length of the back foot also correlates with shoulder height and the combination of these two measures seemed to be more representative than just one. There was no practical way to measure the teeth of these live elephants (without making a plaster cast of teeth while an elephant is under anesthesia), so these other measurements were the next best thing.

We continued to build our ethogram—a list and description of behaviors that fell into particular categories such as affiliative, aggressive, comfort, and reproductive behaviors. We used a behavioral database software called Noldus Observer that allowed us to enter all of our bulls and their behaviors into an electronic data logger instantaneously, as they were observed. This was particularly important when a large group of twelve to fifteen bulls came into the study clearing at once and might stay for several hours. These extended visits allowed us to record quite a repertoire of behaviors for each individual bull.

This was in stark contrast to our previous work with the family groups, which was much more challenging, as they tended to arrive, drink, and leave—visits usually lasting about fifteen minutes and rarely exceeding thirty minutes. We would barely get to observe and log key social interactions other than conflict over access to the best water before the whole family group would move off again. Not to mention that these visits were normally after dark and in bitter cold conditions.

I appreciated getting to keep the more civilized bull hours of around ten A.M. to six P.M. Sometimes bulls chose to hang around and mingle with the females after dark. And there was always the possibility of a bull in musth actively seeking estrus females, but other than that, it was more like the changing of the guards at the water hole between the daytime boys' club hours and the nighttime family group prime time.

Our new research focus meant temporary trading in of hats, scarves, gloves, and night-vision scope for more sunscreen, videotape, dung maps, and lots and lots patience during long behavioral observation sessions. And in time, a dramatis personae unfolded before us.

The Head That Wears
the Crown

———— ▼▲▼▲▼▲▼▲▼ ————

When two related families arrive at a water hole, it's like the front door opening up during a family reunion—little ones squeezing between legs and running off to the rec room in pursuit of a boisterous romp, while the adults stand around engaging in ritual greetings—the handshakes, touching, the excited chatter and all. On one such occasion, I was struggling with the settings on my camera, trying to capture a particularly spectacular sunset, when I heard a wailing in the distance. It was the telltale bellow of a frustrated young elephant bull, standing well ahead of his family, waiting for a cue to proceed.

The youngster shifted weight between his front legs, turning back to see if there were any signs of movement from the cluster of gray forms standing far to the edge of the southwest clearing. This gangly youth was acting like a schoolboy impatient to meet up with his buddies to test out a new sparring move, while his mother set up an invisible tether that he was hesitant to break. It appeared to take everything he had to stop himself from charging ahead of the group that seemed to be waiting for a signal that it was safe to proceed. In the meantime, I could see another herd approaching from the southeast.

A few other young males huddled together behind the bravest one, tossing dust straight up in the air and over their backs with their trunks, as they waited to proceed and mingle with the other group. Light from the setting sun caught in the airborne sand, making it appear as if the young elephants were settings themselves ablaze.

Watching these young bulls interacting within their families took on a new importance, now that we were focusing our research efforts on male elephant society. I found myself thinking about nature versus nurture. How much of adult elephant behavior was pure temperament, and how much reflected the dynamics of the family in which a calf was raised? Since male elephants are born into tight-knit family groups and grow up with a mom, siblings, aunts, cousins, a grandmother, and maybe even a great- and great-great-grandmother, these young bulls would no doubt bring the social experiences of their previous family life with them into adulthood.

We may never be able to understand fully how character traits and past social experience might influence relationships with other adult bulls, but we wanted to document as much as possible. And in order to understand a particular bull elephant's character, I intended to learn something about his family life before he set off on his own.

In particular, I wanted to understand how a mother or the matriarch might shape the character of an individual bull. It's easy to see how a hierarchical society would be beneficial for a young calf of a high-ranking mother. Alternatively, if mom wasn't high ranking, life might not be so easy. And imagine growing up in a family where your brother was the neighborhood bully. If he happened to take you under his wing, it could be an advantage—especially if the fear of his retribution kept other would-be bullies in line. But what if he bullied you? Then, not only would you have to deal with all the bullies from the outside the herd, you'd have to deal with one from within your own family as well.

For example, a cow that we had been following, Wynona, had a male calf that looked to be between six and eight years of age, and one day I observed her in a standoff with the high-ranking Susan. Wynona and her daughter stood their ground as they were visually inspected by the aggressor, while Wynona's male calf, appeared to be the diplomat, reaching out to Susan in what struck me as a gesture of appeasement.

I wondered how Wynona's character might influence her calf later in his life, outside the security of family. Sometimes this young bull saw his mother in a position of strength, standing up for herself when confronted by an aggressive high-ranking family member. Other times he witnessed her fleeing to avoid confrontation. Which example would stick with him as he grew to maturity? Would he learn how to be assertive from watching his mother's brave actions, or would he be less confident than a bull that grew up in an environment where his mother was always respected?

I was brought back to the present moment by more wailing in the distance from the young elephant bull, and the meaning seemed clear: "Why did mom have to wait so long to decide to greet our relatives and have a drink?" "What is she waiting for?" "Come on, mom!"

I shared a similar frustration: Why did they have to wait until just after sunset to decide to break cover? In the dusk, the remaining light was too low to take pictures or to film without a night-vision scope, making the herds more difficult to count and identify. Much to our frustration, the Mushara females were more comfortable approaching water under the cover of darkness, though in this particular situation that made a certain amount of sense since the other group that was amassing might not have been family at all, and the hesitation might have been a reflection of assessing whether friend or foe was approaching.

Despite the challenges of studying groups arriving at night over the years, we were slowly getting to know specific families and were getting a handle on the dynamics within and among family groups that visited Mushara. We named many family groups by the distinguishing features of their matriarchs. Among those who were only occasional visitors to Mushara, for instance, were Left Hook, Left Tusker, Crooked Tail, and Crumpled Ear, while leading the more frequently visiting groups—with which we therefore were quite familiar—were Bent Ear, Slit Ear, Big Momma, and Queen, the head of Wynona's family.

Despite the pandemonium of three, four, and sometimes five herds arriving at the same time, totaling to over two hundred elephants in a few cases, a clear pecking order quickly becomes apparent among the family groups. For example, on a previous night, three groups arrived at the water hole at the same time—the Left Hook, Left Tusker, and Crooked Tail families. Left Hook's family was first to the trough but was quickly displaced by Left Tusker and her herd, which occupied the favored spot at the head of the trough unchallenged throughout their visit.

When Left Tusker finally departed, Left Hook took back the command of the head of the trough, leaving the Crooked Tail clan displaced all the way around the pan, where they were forced to stand in the southeast clearing, huddled and rumbling until it was their turn to drink. When Left Hook left the scene, the low-ranking Crooked Tails were finally allowed to occupy the head.

Other studies have indicated that there are higher-ranking families within an extended family, implying a lineage bias, or a "queen" ethos. Our observations of character traits, or "personalities," among some of the cows seemed to support the notion that "royalty" might play some role in this matriarchal society.

Since the full moon made it safer and easier to collect dung after dark, we were lucky enough to retrieve a fecal hormone and DNA sample from Left Hook and two other herd members. By collecting fecal DNA from these different herds, I was hoping to piece together an extended family tree and determine whether close relatives of the dominant female or extended members of the most dominant family group held greater social status as families than those that were equal in age but less related to the most dominant family.

Although we expected to find a hierarchy within the family as well, we weren't sure whether the hierarchy was determined by strict family lines or by age in this population. Since other researchers had established that the matriarch position was inherited by the next oldest, it stood to reason that hierarchies among individuals in the herd would also be age-based rather than lineage-based, although we wouldn't know for sure until we'd analyzed the samples.

While June was crazy, with so many groups arriving en masse in a chaotic jumble, and usually after dark, by July, group arrivals were more spaced out, with discrete families coming in on their own. This enabled us to do group compositions, noting how many adults, three-quarters, halves, quarters, and babies were in the group as well as sexes of each individual. If we were lucky, we'd also document any dominance interactions between the adult females within the family.

After some careful observation, it was clear that males weren't the only bullies. We had witnessed distinct signs of bullying among female family members—a poignant example of this being the treatment of old Crumpled Ear. On several occasions, we had witnessed a younger cow in the family push Crumpled Ear away from the water and force her to stand at the side of the pan while the rest of the family drank. When this occurred, the family's lack of affiliative gestures and their body language indicated they took no notice of her. It was clear that she was an adult cow with several three-quarter siblings and/or cousins below her in age, yet she was way down in the pecking order.

Crumpled Ear's visible deformity—the complete lack of cartilage in her left ear—might have been only part of her handicap. She may well have had a hearing impairment as a result or some other defect that might have affected her ability to interact normally.

The prevailing understanding among scientists was that once a matriarch became a matriarch, she remained so for life. But practically speaking, it made sense that there would come a time when she was no longer capable of being the group's leader, which entailed making good decisions about where to forage, when it was safe to approach a watering hole, and when there was enough distance between the herd and lions.

I started to wonder about these things while observing a very elderly cow, Broken Ear, who, although the largest individual, was clearly not in command. Several younger female elephants appeared far more vigilant and decisive, and the rest of the family appeared to take their cues from a de facto matriarch and not from Broken Ear. If the oldest member of the group was always the matriarch, then why wasn't this cow still in command?

Broken Ear had a hard time keeping up with the rest of the group and seemed socially out of touch, as we inferred from the way the other elephants treated her. On arrival at the water hole, Broken Ear didn't seem to know the correct protocol of where to go and was gently pushed to the sidelines.

It quickly became clear that not only was Broken Ear not dominant but she also didn't appear to be completely engaged with her surroundings. And this would make sense, after all, considering that elephants are long-lived, intelligent social animals just like humans. Something akin to senile dementia might also plague their elders.

None of the younger elephants were particularly aggressive toward Broken Ear, as compared with what I had seen with Crumpled Ear. They appeared more cautious in how they dealt with her, giving her distance or, at most, a gentle nudge. I wondered what had happened to this elephant, whether she had indeed been in command, and what chain of events might have led to another cow taking over.

We did have the opportunity to see this transition in real time over the period of about five years with a matriarch named Knob Nose, a regular resident of the neighboring water hole, Kameeldoring. Over time, as Knob

Nose became more and more frail, the next most dominant family member, Donut, took over the leadership role.

Rank aside, mothering strategies also vary widely, which would also have an impact on a calf's social experience. During one of the chaotic water hole visits in 2005, I happened to see a baby bull from the Crooked Tail family fall into the trough. It was amazing to see all of the volunteers ready to wrap a trunk around the little one and lift him out of the water.

Curiously, the mother appeared to want nothing of their proffered aid and swatted all the help away. She stood watching over her baby as it swam down the trough, trunk held high, as if she were giving her youngster its first swimming lesson while the others looked anxiously on, eager to step in.

Finally, when the baby got to the end of the trough and was attempting to make its own exit out the ramp, mom engaged her trunk to assist in the final momentum needed to push his little rump out of the water. He then ran to an older sister or young aunt, seemingly to seek comfort. We watched as the relative tucked him under her bosom, holding him with her trunk, as if to reassure him.

Though I could never be sure that I had correctly interpreted what I had witnessed, I at first thought that this mother was cruel, then thought she might have been confused in the chaos of the panic, and finally thought that perhaps she was indeed teaching a lesson at that point. Aren't we all influenced by our mother's lessons of independence?

In a similar rescue in Big Momma's group sometime later in the season, three adult females immediately knelt down together and lifted the baby out with their trunks. This noticeably different and coordinated rescue tactic might perhaps lead to a dramatically different character outcome for the males in the family. It was going to be interesting to see if the calf left to sink or swim grew up into a more self-reliant bull.

Animal character—such as self-reliance—has been termed "personality" by researchers as early as the 1970s. In the 1990s, personality studies on animals were modeled after those done on humans, using what is called the five-factor model, placing individuals within categories broadly defined as negative, agreeable, enthusiastic, open, and conscientious. I hoped to do the same within my studies as well as looking to see if an individual bull's personality could be predicted by those of his female family members. I also hoped to learn if dominance fell into a particular personality formula.

I wished I could have turned back the clock to know what upbringing Kevin—the rival bull in chapter 1 who had failed in his attempt to displace the reigning don—had been like. Did Kevin become a bully due to a negative social environment, or perhaps an environment that was *too* agreeable? And more important, did Greg climb to the top of the bull elephant hierarchy as a result of having had more positive than negative previous experiences or vice versa or an even mix of both with the outcome of being politically astute enough to negotiate the best rank?

Would a bull raised in a dominant family end up in a high-ranking position within his bachelor group? Our findings might well answer the question of whether princes always grow up to be king.

Introduction
to the Boys' Club

———— ▼▲▼▲▼▲▼▲▼ ————

In the orange glow of sunset one evening, I watched a young bull approach from the northeast with trunk swinging loose as a slinky. The baggy skin around his knees and shuffling gait as he neared the water gave him the look of a kid in unlaced, high-top sneakers, with baggy jeans dragging on the ground. Judging from his size, this preteen seemed too young to be on his own, and yet there he was, seemingly unconcerned about not having any company.

Often a bull at this age is just beginning to develop his independence from his family, coming in to drink or departing from a water hole well before or after his relatives. This one for sure didn't seem big enough to have completely severed the apron strings. Occasionally, males have been found to spend time away as early as eight years of age, but generally, they depart from their natal herd between the ages of twelve and fifteen, and this guy looked a little small to be twelve. If he was on his own, it was curious that he hadn't yet found some buddies to keep him company.

When he reached the edge of the pan, I watched as he dipped his trunk into the water, skimming the surface, and then blew out a big burst of water. I assumed that it had been a long time since he had had a drink, as an elephant usually only rinses its trunk when extremely thirsty and needs to clear out dust before drinking.

As this young bull drank, I kept looking up the path, half expecting to see his family appear, while I jotted down some notes about his appearance. His right tusk was slightly higher than the left and broken at the tip. He had a small crescent-shaped notch out of the middle of his left ear. The most striking thing was that his ears were much smaller relative to his body than those of all the other resident elephants, giving him the appearance of a forest elephant from the Congo rather than a savannah elephant.

It's thought that elephants that evolved in a forested environment sport smaller ears because they don't need as much surface area for heat exchange as those that evolved on the more open savannahs of southern and eastern Africa. In addition, the large ears of savannah elephants seem to serve as radio dishes, collecting sound from the environment. Since the thick vegetation scatters and damps sound, larger outer ears, or pinnae, wouldn't be as effective for sound localization in the forest environment—not to mention that such big ears could get in the way of moving around the forest. His unusually small ears immediately brought the name Congo to mind. The name Congo Connor followed shortly thereafter, as this little bull reminded me of my confident little nephew at the skateboard park.

Being as small as he was, I wondered how Congo Connor had managed to avoid falling prey to lions or hyenas. He was of an age that was most vulnerable to lion attack—old enough to slip away from his protective mother and watchful female relations, yet not large enough to protect himself. Since both the lion and hyena populations were on the rise that year, Congo would probably have crossed paths with either or both at some point in his travels. In fact, it looked like some of the hair on his tail had been pulled out, so perhaps he already had had a run-in with a couple of ambitious hyenas, which usually wouldn't be able to do much more damage than that.

As I watched Congo drink, his body language suggested that he was unusually relaxed and sure of himself. The typical young visitor to the watering hole would look over his shoulder frequently, sometimes even mid-sip, letting water run out of his trunk, so he could sample the air, alert either for the arrival of the rest of his family or for some kind of trouble. Yet, Congo appeared unconcerned about his surroundings. What made him so relaxed?

In mapping out his probable past, I assumed that Congo was born on the run, as is often the case. The family forms a circle and waits for a birth, but when the little one sloshes out, there's not much standing around. Within minutes, newborn calves stand and begin to scamper along next to their mothers, ears bobbing and flaccid trunks flopping out in front of them. Within hours, they are suckling.

In the early weeks they gradually gain control of their trunks. The roaring and bellowing that comes from these tiniest elephants seems physically impossible, though it makes sense evolutionarily—just one bellow from a calf left behind during a confused and hurried departure can elicit a rescue squad. Over time, milk tusks appear, then drop out and are replaced by permanent tusks. Young bulls grow tusks faster than do females, which is a useful trait for sexing youngsters at an early age.

Young bulls like Congo grow up among siblings and cousins that seem to delight in each other's company. They spend a lot of time frolicking, dusting, sparring, rolling in the mud, and tackling each other in the water, and they appear to revel in the simple pleasures of being a young social mammal.

Young male calves will sometimes take risks by picking a fight with older males, only to run back to hide under their mother's bellies. Older sisters often take care of these little ones when they get into trouble. Young females will comfort a baby brother after he has been rescued from a mud hole or bullied by another calf.

As a bull gets older, he becomes more assertive and begins to test his boundaries with his mother and sisters. He even tests limits with other animals when they approach the water hole: I've seen young bulls take apparent pleasure in scattering a water hole full of zebra as if they were discovering their inherent might for the first time. They'll even clear the area of guinea fowl if they are so inclined.

When other animals run away from a young bull's challenges like this, his confidence grows, and he can then become even more of a menace. Over time, a young bull starts to get fresh with his sisters and aunts— mock sexual activity known as play mounting. Eventually his mother and aunts get fed up. They may rein him in by giving him a good stiff jab with a tusk, sending him bellowing off to safer territory out of their reach.

But the conflict seldom ends there. Young bulls continue to test boundaries and stir up trouble, including challenging dominant females for access to the best water, until finally, the adult females in the family have had enough. By then, the feeling could be mutual. Over the course of months, a teen bull's need for independence and the family's need for order and peace seem to converge. The young bull takes off. But leaving the

A young male elephant tests his dominance with another species.

comforts and security of family can't be easy. When it's time for a young bull to make his own way, it appears to be a major adjustment.

As I watched Congo drink, I wondered how he might interact with the resident boys' club. He arrived just after some members of the club had left, but not all members had shown up yet, so I figured more bulls were on their way. How would they receive this new male in their seemingly exclusive club?

My questions were about to be answered. Another bull, Billy the Kid, was approaching from the southwest. Billy was another youngster, and a favorite, good-natured punching bag for older members of the boys' club. Billy was even smaller than Congo, and judging by the difference in their shoulder heights, about eight or nine years old. He seemed content with his lot in life, getting to hang out with the big boys, even though they liked to give him the occasional tussle.

As was customary among bulls, as soon as Congo saw Billy, he stopped drinking and turned his back to Billy. I normally saw this behavior when a less-dominant individual was being approached by a dominant bull. Turning one's back to an approaching elephant appeared to be a sign of submission. But Billy the Kid couldn't have been higher ranking than Congo, since he was clearly younger. And rank, at this young age at least, tended to scale with age. Perhaps since this was new territory for Congo, he was acknowledging that he was a newcomer.

Male elephants appear keenly aware of the approach of other elephants. As they seldom fail to check out new arrivals, I was sure Congo was eyeing Billy's every move. Billy, in contrast, seemed unperturbed by Congo's presence, and once he got to the pan, he started drinking from the fresh water outflow.

Congo now turned around and approached Billy. He playfully grabbed Billy's front leg with his trunk—like two boys trying to trip each other in jest. Billy made room for Congo, and they drank together, face-to-face. It was clear at this point that they were at least on cordial terms, if not in fact already friends.

Billy occasionally wagged his tail at Congo, and Congo responded in kind. Congo reached out and placed his trunk in Billy's mouth—a handshake, elephant-style. He then lifted his trunk and placed it over Billy's head—something older bulls seemed to enjoy doing to the younger ones, a sign that they were indeed buddies of a sort.

I knew, however, that Billy's arrival meant that some of his older consorts would be fast on his heels. Sure enough, Tim and Luke Skywalker paraded in just as Congo had placed his trunk over Billy's head again.

Congo dropped his trunk and faced the newcomers with his head up and ears held out, an aggressive posture. Billy stopped drinking and stood sucking his trunk, as if anticipating some kind of trouble.

I hadn't seen Tim since his skirmish with Prince Charles a few days earlier, when Tim had challenged the higher-ranking elephant with a head-on blow. At the time, I couldn't believe his courage. Charles was the bigger bull and he chased Tim several times around the clearing before they disappeared. I wasn't sure of the outcome, but I was secretly rooting for Tim because Prince Charles was another bully, and Tim generally seemed to be the object of his bullying.

The skirmish didn't seem to have ended well for Tim. As he approached the pan, I could see he was pretty beat up, with white tusking scars visible on his face and along his ribs. I felt badly for him. I wasn't too worried about how he would handle Congo, however, as Tim was usually gentle with younger bulls.

But I was concerned about what Luke might do. He was unpredictable. He seemed to be unusually aggressive when putting lower-ranking bulls in their place. Luke's sparring matches were particularly intense—the kind that could easily escalate to serious injury. Although he was missing his right tusk, he frequently won contests against bulls with two full tusks.

Congo stood at the outflow, turning his head back and forth with ears held out, scanning for sounds. When the approaching bulls got close enough, he walked out to greet them, with a slight wag of the tail. He stopped, turned sideways, and looked tentative. Did he notice something about the posture in these bulls as they approached? Was he having second thoughts about how to greet them?

Finally, Congo turned his back to them, the same position of submission he had used when Billy arrived. I could tell that he was a little anxious, as he gave a slight toss of his foot in Billy's direction. Mudslinging always went downhill in the bull world, and older bulls often placed a younger bull between themselves and their adversary as a buffer against any excess aggression. This is why Billy became a punching bag at times.

Luke's body language made me suspect he was looking for a fight. He held his head high and opened his mouth wide in a show of aggression. The fact that he had a stream of liquid oozing from his temples suggested he was coming into musth. (During musth, the temporal glands become enlarged and secrete a thick pungent fluid filled with biocompounds.) Things didn't look good for poor Congo.

The tension eased a bit as Tim gave Billy a double trunk wrap. In this greeting, each bull wraps trunks with the other and puts his trunk in the other's mouth at the same time. Then, Tim gave Billy a slight shove and pat with his trunk, kind of like messing up a younger brother's hair.

Congo watched and waited for his turn to greet Tim. He held his trunk out tentatively, but to my surprise Tim backed up and gave Congo an aggressive head shake, as if to say: "Don't even think about it, pipsqueak!"

Failing at that interaction, Congo then faced Luke who was still giving him his open mouthed threat. Luke put his trunk over Billy's head and gave him a big shove, bigger than usual, and Billy put his trunk in Luke's mouth, as if to appease the tough guy. Billy put up with a lot from these older fellows, but he didn't seem to mind. He let Luke beat on him some more and absorbed a trunk slap, while Congo sidled up to the other side of Billy, rubbing up against him. Poor Billy was caught in the middle again.

Luke and Tim drank at the outflow, while Congo pressed his head on Billy's behind, as if calculating his next move in the shadow of their punching bag. Luke and Tim kept eyeing Congo, to the point where the young bull decided to retreat to the edge of the pan. Things hadn't turned out in his favor thus far.

After a good long drink, Tim seemed to have a change of heart. He walked toward Congo as if to invite him to the outflow and then returned to drinking with Luke. Congo stood there sucking his trunk, seeming uncertain as to whether to take Tim up on his invitation.

Several minutes went by, and finally, Congo decided to make a move. He slowly approached—pushing first on Billy who was between them. Billy put his trunk in Congo's mouth, as if to reassure him. Congo then approached Luke who immediately stopped drinking, raised his head high and held his ears out in a threatening posture. It was as if he was turning to his buddies and saying: "Are you kidding me? Is this guy for real?"

Congo thought better of his strategy and approached Tim instead. Surprisingly, Tim also gave Congo an open-mouth threat. Congo persisted and offered an outreached trunk to Tim.

Tim stepped back. Now Tim and Luke had both stopped drinking and were bearing down on Congo. Billy resorted to sucking his trunk again.

I hated the thought of these bulls coming to blows. I found myself wishing Congo would give up and leave. He appeared to have had enough to drink in the last twenty minutes since his arrival. He could afford to push off.

A few tense moments later, Congo, intentional or not, dropped his penis, which to another bull means he's not going to be a challenge (whether the penis show is really a betrayal of his wary psychological state or a general peace offering is not known). Penis signaling is a very important aspect of how elephant bulls communicate or, at least, how they read each other's mood. Since the elephant penis is almost the size of its trunk and is sheathed in muscle tissue that enables the bull to move it like a limb, it can serve as either a sign of submission or a fairly formidable threat.

No sooner had Congo dropped his penis, however, than there was a rush of blood into this giant appendage, and he started flexing it against his belly—a sign of extreme agitation. It was as if he were rolling up his sleeves for a fist fight.

Congo approached the two larger bulls, penis flexing and all. He was going to take these guys on. Judging by their size, they were at least ten years older than Congo. I couldn't believe his audacity. He placed his trunk in Tim's mouth and then in Luke's mouth and miraculously, just as I expected all hell to break loose, they accepted his greeting. This broke the tension and Billy came over to get in on the action.

Congo rubbed against Billy's flank and Billy on Luke, and they all had a great drink at the outflow together as if they were old buddies and none of the aggressive posturing I'd observed had happened. In a further form of bonding, Luke shoved Billy, and Tim put his trunk over Billy's head while Congo and Luke inspected each other's temporal glands. It was as if they were feeling each other's biceps, testing each other's strength, and assessing their hormone levels. Congo then inspected Tim's temporal glands as well and finished by putting his trunk in his own mouth. This was likely to further assess Tim's hormone level using a sensory structure at the back of the mouth called the vomeronasal organ—a behavior called flehman.

The four bulls proceeded to engage in a gentle sparring match—Luke with Billy and Congo with Tim—that lasted for some time before Congo suddenly decided that it was time to leave. He abruptly ended his sparring with Tim and had one last drink before making his way out, giving a little roll of his head, ears flattened and ready for a decisive departure as he walked away from the watering hole with his nonchalant, floppy gait, trunk swinging out and around like a slinky again.

Although Congo Connor was new to our study, I wondered whether he was new to these elephants. Given the cordial way the encounter had ended, it seemed he was already known to Luke, Tim, and Billy the Kid. I hoped that maybe he'd linger a while in this area so I could observe him the following season. Whatever his decision, Congo had left on his own, and that still puzzled me. Clearly he had the "right stuff" to hang with these fellows and he had taken a huge risk to prove it. Why would he prefer to be alone?

Perhaps it was still early days. If he had only recently parted ways with his family and was just learning to make his own way in the world, Congo might not have had the chance to interact much with adult bulls. That would explain how uncertain and tentative he had seemed at first toward Luke and Tim.

Most bulls are thought to settle far from where there were born, presumably to reduce the chance of mating with females from their own genetic line. This tended to be the behavior observed of young bulls within the park and elsewhere.

I watched Congo head southwest and wondered whether I would see him again. I wanted to find out more about his character, his confidence, and perhaps even which family he came from. But I knew better than to get my hopes up. There was a good chance that this guy might establish his territory elsewhere or get taken by a predator before he could settle on new turf.

Congo Connor smelled the ground with the tip of his trunk and then turned west into the setting sun. As he reached the tree line, the orange horizon swallowed him whole. I would have to wait for another season to learn whether he had won himself a place in the boys' club.

Dung Diaries

▼▲▼▲▼▲▼▲▼

You don't need to have a fecal fetish to want to delve knee-deep (à la Laura Dern in Jurassic Park) in dung. There turns out to be a lot of information in these steamy piles of excrement. Steroidal hormones are relatively stable entities that can be extracted from dry fecal matter and analyzed in a laboratory using radioactively labeled substances that indicate the total volume of hormone within a given volume of fecal powder. This technique has been a boon for wildlife biologists and conservationists wanting to understand stress levels that are reflected in measurements of the hormone cortisol within a given population or, in our case, testosterone as well as cortisol levels.

Fecal hormone analysis turned out to play a critical role in our understanding of how to piece together the saga of the boys' club. It provided a blow-by-blow hormonal account of how they experience their social and physical environment on a day-by-day basis. Consequently, I enlisted a long line of enthusiastic assistants to manage the "dung diaries."

It didn't take long, however, to realize that we needed a full-time dung mapper since there were many bulls visiting for long periods of the day. Because we couldn't get out to the water hole to collect samples until all of the bulls had left, mapping each dung pile and keeping track of which pile belonged to what elephant walking down which path became particularly challenging. It was also important to include information about how many boluses dropped, whether any of the boluses were broken, and whether urination also took place. These details were important if two bulls defecated in very close proximity.

The perspective from the tower was not the same as that on the ground where the elephant actually deposited the sample, so we had to leave one person in the tower with a walkie-talkie to guide us to the correct sample. Whenever we were in doubt, we had to leave that sample behind and devise better mapping skills for the future.

Marking the beginning and middle of each of the heavily trafficked elephant paths, by lining them with a pattern of logs, bones, and skulls was also helpful. And these paths were given appropriate names such as Bolus Boulevard, Cak Cak Court, Dung Ball Alley, and so forth. But our markers along the paths required constant maintenance, as inevitably certain elephants would become curious about these piles and would often dismantle them.

The treatment of the dung, once collected with latex gloves and deposited into a labeled paper bag, took several forms. First, the boluses in the fresh sample were mixed evenly, as there was sometimes a substantial difference in hormone levels between boluses. Given that an individual elephant might defecate anywhere from one to seven or so boluses at a time, there was a lot of potential variance. Then a subsample of this material was bagged, brought back to camp, and placed in our handy twelve-volt solar dung dryer, containing three shelves with four baskets per shelf. We could thus dry twelve samples in one twenty-four-hour period.

Once dried, the samples were brought to Okaukuejo, heated for thirty minutes in a drying oven, and sifted with a colander. The resulting fecal powder was then transferred into a five-millimeter collection tube. Back in the United States, we extracted both testosterone and cortisol levels from each sample as a measure of aggression, stress, and musth status throughout the season.

The samples collected for genetic analysis were prepared differently. For these, we collected multiple small, quarter-sized samples from the outer edge of the shaded side of the bolus, where there tended to be some mucus containing epithelial cells from the lining of the colon. We collected from the shady side, since ultraviolet rays from the sun quickly break down DNA, and we wanted to ensure the greatest likelihood that our samples would yield genetic identifications. The samples were then placed in a salt solution in a sealed sterile collection tube to preserve the DNA contained in the epithelial cells.

After the field season ended, when we had hormone and genetic data, as well as data on associations and behavior of particular individuals over the course of a whole season, we could start to fit pieces of the Mushara male elephant puzzle together. Because it takes twenty-four hours for relevant hormones to appear in the dung, the potential hormonal influence on behaviors witnessed on one day would only show up in fecal samples collected on the following day.

On analysis, we found that high testosterone levels correlated well with the physical and behavioral characteristics of musth, where behaviors previously mentioned such as trunk curling and urine dribbling were present, as well as temporal gland secretions. Urine dribbling had been described in other studies as being the best indicator of musth, and this looked to be the case in the Mushara population as well.

In general, we found that cortisol levels didn't correlate either positively or negatively with rank within the boys' club in the 2005 season. Thus, apparently, it wasn't stressful to stay at the top of the hierarchy, nor did cortisol levels indicate that it was stressful to be at the bottom of the heap, the reluctant recipient of all the displaced aggression from the midranking bulls.

Interestingly, high cortisol levels appeared only after such discrete stressful events as Greg threatening the midranking Torn Trunk, who had been showing signs of coming into musth with his telltale crusty penis caused by urine dribbling and secretions from his temporal glands staining his cheeks. Despite being his best buddy, Greg had Torn Trunk "pinned" (psychologically speaking) at the side of the trough with penis erect for about an hour one day. After this precipitous event, Torn Trunk broke away from the boys' club and fell out of musth. We were able to track this event hormonally after the fact.

Torn Trunk had average levels of around 35–40 nanograms (ng) of cortisol per gram (g) of fecal matter on any given day, but the day after the confrontation with Greg, his cortisol level shot up to 70 ng/g. And, indeed, his testosterone profile matched the physical signs of musth that we observed. They had been below the population baseline of about 250 ng/g and gradually crept up toward the musth threshold of 350 ng/g, but after the confrontation with Greg, his levels returned to baseline and the physical signs of musth gradually disappeared.

Normally, testosterone would increase over a period of time, with musth levels maintained over a period of weeks to several months (for a middle-aged bull like Torn Trunk at least a month or more). Since Torn Trunk was showing the physical signs of coming into musth with testosterone levels on the rise prior to this encounter, and showed no sign of physical injury (which would be another reason why he could fall out of musth precipitously), I reasoned that the encounter with Greg caused him to fall out of musth.

Also interesting is that Greg's testosterone level spiked around this time to 1,000 ng/g (from his average of around 200) and then went right back down to baseline, while his cortisol levels remained constant. Was it possible that Greg was able to manage testosterone in such a way that he only exhibited a spike of this costly hormone within an aggressive event? Whether the aggressive action caused the testosterone spike or the testosterone spike caused the aggression was still an open question, but with evidence pointing to aggression begetting testosterone, not the other way around.

After the stunning incident between Greg and Kevin, when Greg defeated Kevin, his contemporary, who we believed to be in full musth, we were able to show hormonally that indeed Kevin *was* in musth and that he fell out of musth after his encounter with Greg. On the day of the challenge, Kevin's testosterone levels were up in the range of 700 ng/g. Subsequently, we were able to collect four more data points on Kevin in that month and by the third sample, his testosterone was in the range of 100 ng/g, well below the 250 ng/g population average.

Since we didn't have a hormone profile for Kevin prior to June 9, we didn't know if Kevin would have merely been coming out of musth anyway, regardless of his interaction with Greg, or whether the interaction with Greg forced him out of musth. We needed to collect more data on these incidents to be able to tell for sure what was going on. Either way, with testosterone levels in the range of 700, the textbooks say that Kevin should have been able to beat Greg in a contest.

These are just a few examples of the fine-scale resolution data that we were able to get from fecal hormone assays. The other noticeable pattern that jumped out at me from testosterone graphs was that belonging to Tyler. His testosterone levels oscillated throughout the month, as would have been predicted by his erratic behavior within the boys' club, and explained why the elders were on to his mischievous antics and made sure to keep him in check.

Because we found significant linearity in our bull hierarchy in 2005, we were able to test for a relationship between rank and testosterone levels and found no relationship. In fact, it appeared that the more dominant bulls maintained very low levels of testosterone. This followed the predictions of Robert Sapolsky in his ongoing olive baboon study, where he found that during periods of social stability (i.e., when there was relatively little change in the dominance hierarchy), the highest-ranking males did not have the highest testosterone levels. It was only during periods of social instability, after the alpha male was deposed, that there was a positive correlation between testosterone and rank.

Elevated hormone levels correlated well with discrete dominance interactions in 2005. I needed to increase our sample size in the following seasons in order to prove that the patterns were real. I also hoped to be able to determine which came first, the hormone spike or the behavior that resulted in a hormone spike. Although I knew this question would be better answered in a controlled lab environment with mice and blood samples, I was hoping to learn something inherent to elephant dynamics using this approach since such control over the environment, both social and physical, wasn't possible in the wild.

I also wanted to track musth cycles throughout the field season, to understand the waxing and waning of this hormonal state and any potential social influences over specific individual patterns. Was Greg able to suppress some members of the boys' club from going into musth or force others out of musth?

Fecal hormone analysis also made it possible for us to follow the impact of discrete agonistic behaviors toward a subordinate individual by monitoring cortisol levels before, during, and after the interaction. Another thing that I wanted to keep an eye out for was the relationship between cortisol and acts of reconciliation.

In other species, particularly in primates, reconciliation has been exhibited after a discrete aggressive act. In elephants, I had witnessed similar behavior and wanted to see if cortisol levels dropped after situations in which reconciliation appeared to have occurred versus events where there was no attempt at reconciliation.

In these situations, after an aggressive act, such as a shove away from the best drinking position, the aggressor might place his trunk in the mouth of the one that was displaced, almost as an apology. The act would often begin with the aggressor touching his own mouth with his trunk first, as if uncertain as to whether to apologize or let the transgression slide. More often than not, we found that the subordinate would try to reconcile despite what the aggressor did.

These hormone studies turned into an invaluable tool in our research. Thus the dung diaries became an essential component of Mushara camp culture.

Teenage Wasteland

▬▬▬ ▼▲▼▲▼▲▼▲▼ ▬▬▬

Tyler was living in adolescent hell. Just like a teenage boy struggling with a changing voice and swelling Adam's apple, Tyler was an adolescent bull come of age, and the growing pains of his testosterone surges were wreaking havoc on the peace within the boys' club. He had become an unholy terror.

I had first noticed Tyler's testosterone-induced condition because some of the more tolerant older members of the boys' club had been quick to give him a tusk jab on occasion, which I thought unusual at the time. But when I put Tyler's hormone profile together, it became abundantly clear what had been happening.

When I compared the lab results of our dung data to my field notes for the same days, a pattern emerged. On the dates that I noticed bulls like Abe giving Tyler a jab that seemed uncharacteristically severe, fecal hormone analysis showed that Tyler's testosterone levels had been high and then dropped after the slap down from an elder. He must have stepped over a line and been slapped back down, though I hadn't seen the infraction, just the punishment.

I had to return to some of the video footage and watch it frame by frame to see that indeed Tyler had been the agitator, although in ways so subtle I had missed them at the time. The video caught him being the instigator with the other bulls on a number of occasions.

Tyler would try to pick a fight by initiating a sparring match with a shove rather than a gentler invitation with a trunk grab or a trunk out-stretched. Low-key sparring was the male elephant equivalent of a friendly arm-wrestling match. If an invitation was ignored or rebuffed, Tyler wouldn't give up. He kept shoving members of his cohort, leaning his head into their side, trying to trigger a reaction. That's when Abe had stepped in and sorted things out with a quick shove or head butt.

As I replayed the tape, I thought back and tried to figure out how I'd missed all this. At the time Tyler's antics had unfolded, I'd been paying more attention to the bulls that were slightly older than Tyler, like Luke, Tim and Prince Charles, where I expected to see such pestering.

Bulls in their early teens that still lived within their matriarchal family groups really enjoyed pummeling each other. And not only that, they also enjoyed mounting each other and seeing just how much they could get away with, both with males and with females.

There was a definite difference between the rambunctious behavior of the young bulls still within the family circle and the behavior of young bulls newly moved into all-male society. Our behavior data and hormone analysis suggested a possible explanation. It seemed as if the older bulls were keeping the youngsters in line by suppressing their erratic coming-of-age testosterone spikes.

Socially induced hormonal suppression is a well-known phenomenon in the animal kingdom. For example, in the mandrill, a close relative of the baboon, secondary sexual characters do not develop in subordinate males. "Fatted" males, as they are called, are social, with highly developed genital skin coloration, large testes, high testosterone levels, and fat rumps. "Non-fatted" males are more solitary, with pale skin, low testosterone levels, and more slender rumps. And while there appears to be a gradient of testosterone levels from the most dominant to most subordinate individual, there also seems to be some threshold at which only the alpha males with the highest levels of testosterone become these "super" males.

In orangutans, studies indicate that sexually maturing adolescent males have a significantly higher level of stress-hormones than younger or older males, which suggests that stress hormones might be suppressing secondary sexual characters in some male orangutans. The opposite could actually be the case, however—non-fatted males might suppress their sexual displays to reduce their own stress levels. Thus the lack of secondary sexual development could be an adaptation, the purpose of which is to avoid stress (and potential attacks by dominant individuals) during adolescence.

This phenomenon has also been hypothesized in killer whales, where dominant males are much larger, with larger dorsal fins than subordinate males. Researchers have also reported evidence of sexual suppression of subordinate males among social lemurs, meerkats, and many other species. Suppressing sex characteristics to avoid conflicts isn't without its drawbacks, however. Some investigators have suggested that growth factors and immunity can be suppressed in subordinate male baboons, leaving them smaller and sicker, which would then lead them to be more vulnerable to predation.

Is it possible that teenage boys might experience suppressed levels of testosterone in the presence of adult men? Since this phenomenon is known from nonhuman primates, it's intriguing to wonder if some similar phenomenon operates around boardroom tables, pro-wrestling matches, and boot camps.

Studies with elephants in South Africa have also shown that the presence of older bulls serves to manage the younger bulls' expression of testosterone and, thus, their patterns of aggression, just as we had witnessed with Tyler. Taken out of their natural environment among older bulls and left to their own devices, young bull elephants will become hell-raisers— the equivalent of juvenile delinquents.

One dramatic example of this phenomenon took place in a private reserve in Pilansberg, South Africa, between 1992 and 1997. Young bulls less than ten years of age were relocated to this park from their natural social environment after a management decision was made to cull in Kruger National Park in the 1980s. These young bulls entered their first period of musth at a much earlier age than those studied in intact social communities. Successful breeding began at the age of eighteen, when they were observed to experience initial musth, rather than around their midtwenties, which would be expected in a normal population.

These young bulls had been introduced into an area that was also home to a population of highly valued white rhinos relocated from elsewhere in South Africa. The stage was set for conflict, as elephants and rhinos are natural antagonists. Though smaller, rhinos are stubborn, and they tend not to back down when challenged by elephants. This refusal to back down appears to provoke elephants greatly.

Into this already charged atmosphere came a group of unsupervised teenaged elephants that quickly went into premature musth. Unchallenged by higher-ranking males, these younger elephants had testosterone levels up to fifty times the normal level. They actually went berserk. When confronted by cantankerous rhinos, these bulls ended up attacking and killing forty rhinos before it was decided that something had to be done to stop them.

Out of desperation, managers at the preserve decided to introduce six older bulls to see if they might somehow hold these teenagers in check. Researchers had been monitoring the hormone levels of the younger bulls and could therefore compare what happened before and after the introduction of the older bulls.

Sure enough, on the arrival of the older males, the younger ones instantly fell out of musth and stopped their delinquent behavior, thus providing clear evidence that the mere presence of older bulls acted to suppress the expression of testosterone in young bulls. And to strengthen the veracity of this pattern further, the phenomenon actually occurred in a second reserve in South Africa as well and was resolved in exactly the same way.

Although these two extreme elephant-management problems in the parks were corrected by relying on hormonal suppression, it also suggested to me that young bulls might benefit from good male role models. That is, was something more than hormonal suppression at work? Did these younger bulls need social guidance and boundaries that older bulls provided in order for them to remain decent members of their society? A further exploration of mentoring and mentorship could provide valuable lessons for the handling of captive and reintroduced elephant bulls. It could also well be that humans might learn much from understanding how nonhuman societies manage these vital interactions.

The occurrence of such violent outbursts and the need for mentoring raises a host of secondary questions. What role does displaced aggression play in all-male societies, and is there an adaptive advantage to bullying? Studies on rats in stressful situations have revealed that rats allowed to bully another rat had lower levels of stress hormones than rats that did not have bullying as an outlet. Elephants may well use displaced aggression for a similar purpose. Would the levels of stress hormone of aggressive bulls be lower than those of lower-ranking individuals that were consistently nudged, poked, swatted, and shoved?

The only way to prove that the behavior of the young bulls in Pilansberg was caused by their traumatic past would be to relocate young bulls that had recently left their families under normal conditions and compare the two groups. Such data were not available, but I was hoping that my research would provide a natural laboratory to obtain long-term, high-resolution hormone data on known individuals of different ages in order to better understand what was witnessed in Pilansberg. These studies might provide additional insights into the sometimes inordinate bouts of aggression exhibited by young elephant bulls at Mushara and elsewhere in Africa, bouts that tend to be tempered by their elders.

As I watched Tyler struggling with his hormonal surges, I was reminded of a recent study that attempted to link juvenile delinquency and early exposure to alcohol to risky decision making as adults. In the study, one group of young male mice (at ages equivalent to human adolescence) were dosed every day for a ten-day period with ethanol embedded in a sugary gel called Jell-O shots. A control group was fed a gel without alcohol.

After the ten-day period, with alcohol-conditioned mice ready to go, the scientists trained each group of mice to select one of two levers, the one on the left would produce a treat of three sugar cubes every time the lever was pressed. The lever on the right would sometimes result in four sugar cubes or no cubes. In a fascinating set of experiments, those mice exposed to alcohol in their youth consistently exhibited risky behavior as adults. Time and again, the alcohol-induced mind couldn't resist the Vegas black-jack-table response—the all-or-nothing lever. And worse, when the odds of getting a treat were reduced over time, the high rollers continued to make a risky choice, resulting in many fewer treats in the end.

This begs the question of whether a young bull entering musth prematurely, in the absence of elders, might exhibit risky behaviors later in life. To my knowledge, no one had yet studied the effects of such experimental exposure to high levels of testosterone at a young age with behavior later in life. But it did make me wonder about Tyler and other young elephant hell-raisers and what the future held for them.

As the 2005 season came to a close and August neared, it got windier and windier. It was like that in the dry years. And what made it worse was the lack of grass and scrubs to help hold the sand in place when the wind picked up.

The dust devils were increasing in frequency and magnitude. The whole camp had to duck and cover (covering as much of the electronic equipment as possible) when a mini-twister came through—a wall carrying half the clearing's worth of sand, leaves, and elephant dung in its midst. If an unlucky team member had a tent door open during this time, there was no hope of rescuing the sleeping bag. These were gritty times indeed.

Just before the season ended, I watched Tyler holding his own in a long sparring session with the bruiser Luke. I wished I could turn the clock forward to see how these young elephants would turn out—to find the truth behind our elephantine teenage wasteland.

Coalitions and a Fall from Grace

———— ▼▲▼▲▼▲▼▲▼ ————

I woke to a cry in the night. Lying still, I held my breath and waited for more. Nothing. Only silence. Without making a noise, I lifted my head to hear with both ears. I pulled my fleece hood down and exhaled. It seemed as if the cry had come from right below my tent platform, just next to camp, but I had been in such a deep sleep, I couldn't be sure. Ever since my husband, Tim, left Mushara earlier in the season, I sometimes felt I was in a world of my own inside my tent. Just me and the African night—and the predators.

I listened for the rustling of sleeping bags from other tents, wondering if any of my research assistants had heard the noise. I heard nothing but the sound of my own pulse throbbing in my ears.

I looked out and despite the brilliant stars, the night was pitch-black and I couldn't see a thing in front of me. I was exhausted. There had been so much activity that night that it was hard for me sleep, let alone want to fall asleep with so much sound to record. So, when I finally did get to sleep, it was more like passing out.

Before long it came again—the noise that had jolted me awake. First there was a sickly eerie squeal at full volume, more unsettling than nails on a chalkboard. I winced at this unholy noise. I wished that I had kept the side panel of the tent open so that I could have looked down to see what was happening on the ground outside the camp perimeter. But before turning in, I had rolled down and Velcroed the side flap to keep out the cold. And I couldn't reach my head out the front to look for fear of making a noise.

There came a high-pitched whine again, and finally the same ear-piercing sound that had awakened me. A strange resonant growling accompanied the hideous cries from all sides, presumably from the aggressors

Finally, came the telltale whooping. I was sure now that it was a hyena in trouble, and since there was plenty of noise to cover any sound I might make, I slowly moved my hand next to me to reach for the night-vision scope. I pushed myself forward onto my elbows to clear the tent flap and have a look.

There, next to my tent just feet below me was a male hyena (I had surmised by his smaller size), tail and legs tucked under himself with rump to the ground as he seemed to be begging the seven surrounding female pack members for forgiveness for some unknown transgression. The seven closed in, their heads held low and tails all fluffed up and threatening. I was amazed by how much tail signaling went on in a species that, on most occasions, shows not even a sign of having a tail.

The poor hyena's face contorted into a terrible grimace as the pathetic creature emitted the most horrific screeching cries again, made worse because I could see his teeth and eyes glowing a satanic green through my night-vision view. He crawled a few inches in a supplicating crouched position, feet tucked under belly and then screeched again, starting at a high pitch and holding it there.

And just when I thought I couldn't bear the screeching any longer, he began a low whooping, head facing the ground as he cried out. An accuser approached more closely, growling rhythmically again, then lowered her head down and also began whooping that surreal, signature hyena announcement call, akin to the scaling of an octave. The other hyenas giggled uncontrollably—a contagious demonic-sounding giggling, unparalleled in nature.

This dominance ritual and hideous supplicating went on for some time before the offender was allowed to go free. I couldn't tell what it was that caused the dominant members of the pack to relent, but the young male was finally released from their imposing circle. I wished that I had had the presence of mind to film the interaction, but I was too stricken with the power of the moment to do anything but watch with morbid fascination.

Almost immediately afterward, two lionesses closed in on the scene and the pack scattered further. Perhaps they had smelled the lions' approach, and that is what saved the young male from what looked like certain physical injury.

Lion and hyena populations at Mushara seemed to follow a cyclical pattern of expansion and collapse, accompanied by certain patterns of dominance. In 2006, when I witnessed this particular confrontation, a hyena den was expanding relative to the previous season. As it grew in size and power, it began to show signs of internal strife. While our most recent goal was to understand dominance within the male elephant world, I couldn't help noticing over the years what pecking orders existed within and among other species at Mushara.

How dominance is established and held within a social group can be influenced by the dynamics of a particular group as well as external environmental conditions. And a shift in power might originate from the unexpected rise of a risk taker.

As we got to know the bulls and individuals, for instance, we found particular patterns of behavior. Kevin was the bully, always ready to spar, to initiate a contest that would inevitably turn out badly. Then there was Willy Nelson, the raggedy but well-respected softy, who was constantly being stalked by Kevin as a sparring partner. Jack Nicholson, the sweet talker, got all the others in a congenial mood with his affiliative gestures. Luke Skywalker was the single-tusk sparring champ who put rabble-rousing youngsters in their place. Torn Trunk was the midranking loyal right-hand man to the don, Greg. And then there was the mild-mannered second-ranking Mike, whose rank was a bit of a mystery due to his lack of aggression.

Although we knew that Greg was in charge, the big question was just how long an elephant bull could maintain his reign. Among primates, it typically varies from a couple of months in some species to as long as three years in the olive baboon. The dominant male primate is usually overthrown by being killed or wounded beyond the ability to challenge his successor.

A leader in the human world, when challenged, might surrender peacefully, engage in some sort of truce, flee, or fight to the death. Elephant bulls have been known to engage in mortal combat, so it may also be true for elephants that they remain in power until they are no longer physically fit enough to defend their position. But the length of reign could also be a character-dependent phenomenon, as is the case among some human societies.

Greg had an interesting combination of behaviors that ranged from being tough on bullies to being gentler with the young bulls. He was very solicitous of and patient with the younger bulls when they approached him, always ready for a gentle spar or a trunk over the head. On one occasion, we even witnessed him allowing a younger bull to suck on his tusks, a behavior I had never before seen and one in stark contrast to some other middle-aged bulls who would give a quick jab to a young bull when they'd had enough of their playful invitations to engage in bodily contact. I wasn't sure how unique Greg's strategy was, so I was especially eager to see whether our data would show this pattern to be consistent from year to year.

The extremely dry environmental conditions of 2005 meant there were especially few places to drink in the region. In order to minimize conflict over access to available water, a stable linear dominance hierarchy would need to be clearly in place and was. Nevertheless, there were some unexpected exceptions, as I learned, this time again involving the upwardly mobile Kevin, the third-ranking bull confronting the mild-mannered and second-ranking Mike.

Mike's behavior was particularly unusual given his impressive stature. He had the physique—including perfect and formidably wide-splayed tusks—that would intimidate even the most ardent of bullies. But, for whatever reason, he appeared to choose the passive approach. Over and over again, we watched Mike walk away from conflict rather than respond to a challenge with aggression.

I had to wonder how long this passive strategy was going to last for Mike, if in fact it was a strategy. Based on what we'd seen among other males, it seemed that bulls needed to exhibit a certain amount of aggression to maintain their position in the hierarchy. Having no upwardly mobile aspirations at all could have meant that Mike's rank near the top of the hierarchy was an artifact of strategically avoiding conflict and thus not getting displaced, as he would have been if he were more aggressive. By letting the other bulls fight, he was rising up through their ranks unchallenged. But it was on one fateful day three weeks into the field season that we were to see how this strategy finally played out for mild-mannered Mike.

The day started out just like any other for the small contingent of the boys' club gathered at the water hole. Kevin, Stoly, Jack, and Abe were having a fairly peaceful drink when I noticed Mike, our gentle giant and second-ranking bull, making his way up from the south along the southwest elephant trail. He sauntered along in his habitual slow gait, but when he got to the edge of the clearing, he stopped short—as if suddenly aware of the presence of an elephant he hadn't expected to find.

Mike stood frozen in place for a long time, seemingly uncertain as to whether to approach the water or wait until the others had departed. This was a strange turn of events, since these four bulls had always been Mike's allies. And even though Kevin was a bully and had clearly been working hard to exhaust Mike over the course of the season, Mike always seemed able to keep him at bay. But perhaps Kevin had finally worn poor Mike down.

As Mike stood there, he suddenly reminded me of the Cowardly Lion in the *Wizard of Oz*, a formidable beast holding his tail (or, in this case, sucking his trunk) and shuddering in fear. Clearly, something had happened to change the social dynamic of the boys' club for him. His confidence appeared completely shattered.

Mike stood cowering at the edge of the clearing for over an hour with his trunk hanging over his tusk. Having been very familiar with his social habits, we couldn't figure out the cause of his hesitancy—his buddies were all drinking, and he seemed more than anxious to join them. But it was as if some mysterious force held him back. We all waited to see what would happen next.

Finally, after much deliberation, Mike made his move. But as he edged toward the other elephants, we saw what had been holding him back. Kevin must have sensed that Mike had been standing at the edge of the clearing. As soon as Mike took a step forward, Kevin stopped drinking and positioned himself at the head of the water hole. He held his head up and ears out—eyes fixed on the intruder. Mike seemed to brace himself for a confrontation but continued forward.

At this point, we witnessed something extraordinary. Kevin's behavior appeared to stimulate a cascade of similar behaviors in the three other large bulls that had previously been Mike's friends and allies. Stoly, Jack, and Abe looked up at Mike and immediately stopped drinking and stepped away from the water. They formed a line with Kevin at the helm. All four elephants now stood with their heads up and ears held out, challenging

Mike. Kevin appeared to have orchestrated this coalition. I couldn't figure out how he had done it, and I could only guess that he was finally attempting to grab power and secure the spot as second in command in the boys' club pecking order—a position Mike had held for the past two seasons. Perhaps after his defeat with Greg, Kevin's new strategy was to come at the hierarchy from the bottom up.

They stood in line, four incredibly large elephants, like a gray brick wall, aggressively staring Mike down. Poor Mike looked as if he might spontaneously defecate in anticipation of the elephant equivalent of a firing squad. Instead, Mike went back to sucking on his trunk, standing, facing the others for some time as if waiting for them to stand down and greet him. But his accusers showed no sign of backing down. Four angry elephants loomed before him, ears spread wide in a show of determination, until Mike eventually relented and slowly headed down to the end of the trough where he drank quietly on his own at the lowest-ranking position.

True to his usual strategy, Mike had chosen not to fight. But this time that strategy resulted in him being toppled from his place in the club's hierarchy.

Up in the tower we shook our heads in amazement. What had happened out there in the bush to generate such tension between these former allies? Clearly, some drama had played itself out somewhere, out of sight of our observation tower. What had we missed? Or was it the testosterone talking? Perhaps displaced aggression after Kevin's fallout with Greg?

Whether under the influence of testosterone or not, none of the other bulls, I was fairly certain, would have behaved this way toward Mike on their own. How had Kevin succeeded in rallying the other three to his side? By forming a coalition, he was now one step away from usurping Greg's position and becoming don himself. But what advantage, if any, did Stoly, Jack, and Abe get in exchange for switching their allegiance from Mike to Kevin?

Male social strategies vary widely, depending on population density, competition for resources, the nature of the threat of predation, and the need for protection, as well as the inherent sociality of the species. Male alliances are well documented in species such as the bottlenose dolphin, lion, horse, baboon, macaque, mountain gorilla, and chimpanzee. They are formed generally for the purpose of defending territories against competing males, as well as for cooperative hunting, increasing rank, and/or facilitating access to females.

Coalition formation is well known in primate societies, where an upwardly mobile individual gathers up some cronies in order to topple the dominant individual in situations where he can't succeed on his own. It would make sense that male elephants would form alliances and even coalitions to defend against a challenging male, to improve rank, or possibly to facilitate better access to females. And Kevin made it clear just how advantageous a coalition could be within the boys' club.

Male tolerance of other males tends to be lower in species where spatial and temporal access to estrus females is at a premium. This is, of course, true for elephants, where the problem of access seems to be managed by the mechanism of rotating musth and taking a "gentlemanly" turn at engaging the girls. These gentlemanly turns, however, may not be so gentlemanly, depending on prior associations, including position within the hierarchy.

But how lasting were the outcomes of coalitions? Was the princely Mike permanently unseated from second place? Did this mean that Kevin would eventually challenge Greg again? He hadn't prevailed when he was in musth at the beginning of the season, but perhaps with some formidable allies in tow, he'd be more successful in the future. The answer to this question was probably related to the motivation level of the instigator as well as the ambition of the ousted individual.

For the rest of the season after the coup, we observed that Mike took even longer to break cover and enter the perimeter of the water hole—well over an hour—and when he finally did, he approached with the slowest gait I had ever seen, almost as if his feet were too heavy to lift. Although it was impossible to know the state of Mike's psyche, I felt certain that his chances of rising to his former spot near the top of the hierarchy were over. I feared that the prince had fallen from grace.

Male Bonding

———— ▼▲▼▲▼▲▼▲▼ ————

While watching what appeared to be a particularly affectionate bout of interactions between Tim, Dave, Jack, Luke, and Torn Trunk one afternoon, it was hard not to think that they were really enjoying themselves. Jack was in a particularly congenial mood when he arrived, greeting the others with a trunk-to-mouth hello, not waiting in a standoff as the midranking young adults sometimes did, as if to see if someone else might initiate first.

After greeting the others, Jack marched up to Luke and wrapped trunks with him in a way that I hadn't seen bulls do—or any elephant for that matter. They stood face to face with trunks entwined in three twists beneath them. By all outward appearances, this was definitely a male bonding moment. But what does male bonding mean, exactly?

Strong male bonds exist in some species despite the pressure of the competition for females, the most striking and well-documented example being the chimpanzee. In chimps, male bonds are thought to exist to minimize conflict or even acts of infanticide by other males within a group or as a mechanism of protecting against outside groups of males accessing resident females. These bonds are described as forms of coordination and mutual attraction.

This is not to say that aggression is nonexistent in bonded male groups, but the interplay between affiliative and agonistic interactions is thought to strengthen bonds—the equivalent of two male friends that find themselves in a heated fistfight over a girl, or even over which sports star is considered more robust, and then making up afterward and having a drink together at the bar.

Thus, affiliative behaviors are used to avoid conflict or appease a buddy after a conflict, while agonistic behaviors are exhibited to attain, defend, and maintain a dominance rank (or that of their favorite sports star, as the case may be). And it is the interplay between affection and aggression that appears to facilitate and maintain healthy bonds.

Ritual is an important aspect of male bonding. In many human all-male societies, this is evidenced in such customs as hazing in fraternities, initiation into traditional gentlemen's clubs, induction into boy scouts, and boot camp initiations. Ritual is also important in religious ceremonies—many of which have been reserved to men only.

Ritualized dominance behaviors within these bonded male groups are thought to formalize status relationships, where acts of subordination toward dominant individuals may serve to preempt aggression and thus generate greater tolerance among group members—hence the ceremonial stoop and kiss of the mafia don's ring to quell any thought of insubordination, disloyalty, or coalition formation against the don.

Although male elephants hadn't been described as forming bonds within extended groups, I was slowly building a case to demonstrate that the boys' club was a group of male elephants that had formed long-term associations and indeed engaged in bonding behaviors. In females, bonding has been defined in the context of tactile and vocal greetings, cohesion, vocal coordination to motivate action, and coalitions. Over the years, we had documented these very things on a consistent basis within male groups such as the boys' club.

The highly ritualized trunk-to-mouth greetings, for example, appear to serve not only as a greeting but also as an act of subordination of a less dominant bull—a signal to the dominant individual that the lower-ranking individual recognizes the other's dominance. The trunk-to-mouth ritual also appears to serve as a solicitation of further affiliative exchanges, most often seen initiated by a youngster to an elder or a youngster to another youngster. This type of ritualized bonding behavior is very similar to that seen in male chimpanzees.

Then there were the coordinated vocalizations on departure initiated by Greg's "let's-go" rumble. These call sequences were similar to chorusing, but with nonoverlapping repeated calls. The let's-go rumbles almost always seemed to be initiated by the dominant individual, who would step away from the water hole, stand still and vocalize, usually while flapping his ears, as is seen with matriarchs rallying the family to depart. Others paid attention to these vocalizations and stopped drinking.

After the dominant bull's first vocalization, a second vocalization would usually follow, most likely from a subordinate. This second vocalization would begin just as the dominant bull's vocalization ended, followed by another such call, and another, in a repeated series. All would eventually follow the dominant individual on to the next venue, while a series of these vocal bouts were repeated along the way. This was definitely a formalized coordinated behavior that was ritualistic in nature and, as far as I had witnessed, only occurred within groups of close associates.

Finally, coalition formation was also clearly evident, as witnessed between Kevin and his cronies against the gentle Mike and, later, between Greg and his right-hand man, Torn Trunk, against the poor old Captain Picard. So the criteria of bonding in female elephants—tactile and vocal greetings, cohesion, vocal coordination, and coalitions—were unmistakably evident among these males. They were indeed bonded.

Why hadn't such a clear case been made for male bonding in other male elephant populations? Associations in male elephants that take the form of bonding may be stronger in the semidesert environment of Etosha National Park, Namibia, where we were observing than in other parts of Africa, where water is less restricted. Inevitably, dry environments, particularly striking in exceptionally dry years, have the effect of compressing elephants, forcing them to drink at the water hole together and to interact more frequently in larger numbers than they otherwise might, given a choice.

This could explain the higher level of associations, quantitatively and qualitatively, seen within the boys' club during the dry year of 2005 as compared with an especially wet 2006. However, there was also a significantly greater level of associations among the boys' club members, in comparison with the resident population as a whole sampled during both wet and dry years, evidence of an active choice regarding with whom to associate, another measure of a bonded group.

When water is limited, it could be more important for males to form friendships and alliances to avoid conflict over water access, and in this environment, the forming of coalitions against other groups of males may thus be important to surviving in a harsh environment. Consequently, sociality in male elephants may be heavily influenced by environment, and in arid environments, male elephants may form closer bonds and longer-term associations than described elsewhere.

It is therefore essential to understand both the dynamics of relationships within an individual social group and the potential influences on those relationships in order to understand the underlying mechanisms that drive the group's inherent structure. There are good long-term data describing the nature and dynamics of male dominance hierarchies in mixed-sex societies such as baboons, mountain gorillas, and chimpanzees, but very little published data on dominance hierarchies and evidence of bonding within male elephants.

Dominance is thought to be based mostly on intrinsic factors relating to age, as well as size, as elephants continue to grow well beyond puberty and, in males, until much later in life. The fact that there is extreme sexual dimorphism in elephants, males being twice as heavy as females and about a third taller, is indicative of a polygamous mating strategy with intense male-male competition in which a male competes with other males to mate with more than one female. Yet dominance hierarchies have only been reported at the level of one-on-one contests in male elephants, revealing little about their social structure. And given their reproductive strategy, bonding would not necessarily be predicted.

However, considering that males spend much of the year outside the state of musth and among other males, a lot of the time they spend together does not involve competing for females. And within these periods, it would seem advantageous to build a buddy system, whether it be to muscle in on the best fruit tree or the purest water or to guard against bullying. And if you think about it, male bonding doesn't have to be any more than a high five or a secret handshake to clear the air of any tensions.

These gestures are akin to the haka—the ancestral Maori war dance—performed by the All Blacks before a rugby match, a battle cry or to the drummer on the sculling team, synchronizing and uniting the team in a common cause. All these things are thought to enhance bonds between individuals and create a kind of collective mind. Perhaps, at a minimum, Greg needed his posse to intimidate potential encroachers on their preferred resources. Or perhaps by garnering a larger group of associates and

suppressing hormones within the ranks, a larger portion of the musth period could be monopolized by the dominant individual.

As I watched Jack and Luke entwine trunks in a secret elephantine
handshake all their own, I applauded their unabashed displays of what
seemed like all-out affection. These bulls were not shy about their public
displays. There indeed was a case for male bonding among bull elephants.

The Domino Effect

▼▲▼▲▼▲▼▲▼

It was about one A.M. when the breach of camp took place. It was now four A.M. as I watched a red three-quarter moon sinking below the horizon. The constellations sat brilliantly against an inky black sky. Scorpio had never looked so prominent as I lay in our double bedroll up on the second floor of the tower, trying to stay awake.

A lion called in the distance. I got up and poured more tea. My head felt the chill through two layers of hats. Only two more hours to go.

I was on watch until sunrise. Tim and I were trading off. With a pride of eighteen lions in the vicinity this year, and plenty of troublesome two-year-old males, sure enough, our luck ran out: one of them had decided test our boma cloth perimeter in the wee hours that morning.

A thirty-year record-high rainfall in 2006 had created very different conditions at our field site, bringing a marked increase in predators. The success of the predators was evident in larger den sizes of both lions and hyenas. And the increase in jackal pups was no doubt supported by supping on an inordinate number of springbok and gemsbok lambs born after the heavy rains.

Pride and den expansions sometimes caused age-old rivalries to flare up into excruciating conflict, in some cases leading to mortal combat. Abandoned cubs and pups were just part of the collateral damage. The rangers reported a whole litter of lion cubs abandoned at a nearby water hole, and we had an unfortunate hyena pup wander into camp, looking for protection after its mother most probably lost a turf battle.

After two nights of howling next to camp, which was protected by the electric fence barrier from lions, the pup finally stumbled off into the bush in search of its destiny. Despite the plea by some team members to be allowed to intervene, the rangers asked that we not do so.

On this particular night, Tim had been awakened by a swatting and tearing sound right next to our tent at about twelve-thirty A.M. I hadn't slept the night before and therefore, after watching the pride settle in at the water hole at ten P.M., I had drawn my head within the sleeping bag to hide from the moon in an attempt to get some sleep. Fortunately Tim woke to the noise.

We both sat up to find a young male lion just below our tent, face and claws fully engaged with our ostensible barrier. He had shattered the illusion of protection, and our sense of safety evaporated.

Tim bellowed. And for a sickening moment, his outburst was met with only silence. Nevertheless, the deep sleep of the camp had been shattered.

The young lion took a step back and stared at us. Three others sat several meters away, providing back up. Some primal instinct—man versus nature—caught hold of Tim and threw him into a rage.

Tim roared and yelled at the lions, as I fumbled around in the tent looking for something he could throw at them. I handed him a water bottle, which landed right next to them with little effect.

Tim opened the bear chaser, the extended-range version that we had brought along for this very purpose. Tim sprayed the tear gas at the lions, but they had stepped just out of range of our deterrent, and with almost no wind, the spray was more noxious to us than to the lions.

After throwing all of our steel tent pegs at them, Tim's voice having gone hoarse from yelling, we ascended the tower to shine a spotlight on them and to make a plan, instructing everyone else to remain in their tents. We then tried playing back a few lion calls through our speaker system to see if the territorial calls of another pride would cause them to flee.

The calls only served to make the pride more curious. As many of the pride members were already lounging about the water hole, they gathered around the speaker with great interest. Tim then resorted to attaching a microphone to the speaker system and yelling at the lions through that. This seemed to provide great amusement not only for the lions but for the rest of the camp as well, the others not yet realizing the gravity of the situation.

We devised a new plan to drive them back. One person would shine the spotlight from the top of the tower on the pride as Tim and I raced at them in the truck, honking the horn while our Japanese volunteer shouted out samurai insults through the speaker.

We managed to push the lions back to the perimeter of the clearing, but not before getting charged by the young males. We were extra cautious in our maneuvers so as to avoid getting surrounded, as there had been several cases of lions attacking vehicles in the park when provoked, even to the point of a lioness jumping through a rear window.

Every year there was a slight variation in the lion scene at Mushara, depending on the weather and the social relationships of the lions themselves. Since Mushara is a fairly isolated water hole, it had generally been assumed that its surrounding territory was too large to be held by a stable pride. In the past, I had only contended with a few transient young males that were fairly skittish, or a pair of honeymooners, or a mother and a few cubs for a several-day period, but we had never had to deal with a whole pride settling in at Mushara for an extended stay.

After contacting Windhoek the next day to organize the delivery of some additional electrical fencing supplies, we engaged Johannes Johannes, the park's research technician, to serve as a hired gun for the next several days until the shipment arrived. We then set about reinforcing our little fortress, building a far more substantial fence than we had originally planned, which did indeed succeed in deterring the curious young lions at the base of the fence. I caught them on several occasions inspecting the bottom strand of wire at night, one young male even touching his nose to the fence, jerking his head back with a grimace and snarl. Although it was disconcerting to see him react as if the electrical shock was no more than a bee sting, the new barrier nevertheless seemed to be effective.

The fence also attracted the attention of a few of the more curious elephant bulls that showed up in the middle of the night—including Luke Skywalker. I watched him sneak up to camp, practically on his tiptoes, to check out our new addition. He then leaned way over and stretched his trunk out to its full length in order to touch the hot wire between the tips of his trunk.

He got such a shock that he released his bowels right then and there.

And for an elephant bull, that's more than a wheelbarrow full of feces. I almost laughed out loud, but then thought better of it. This was his home that we were intruding on, after all, and we had turned it into a hostile environment. I consoled myself with the knowledge that we'd only be here for two months, and then we'd pack up the whole operation, stinging fence and all, and return the site to its rightful owners.

Meanwhile, despite the few curious bulls in the night, it had been a painfully slow start to the season, with hardly any fresh signs of elephant in the area. The elephants had taken much longer to return to their dry-season territory, still having many options for foraging and watering in outlying areas. It was thus an unfortunate year to have arrived on the first of June.

We set up camp with not a single elephant visit, and I started to worry about the expectations of my volunteers. What if the elephants never came?

I was used to the possibility of a season going awry, given the unpredictability of nature, and could easily justify a slow season in a report to a funding agency. But now that I was taking on volunteers who were contributing to the research, there were expectations to be met, especially regarding the presence of wildlife—in particular, elephants. These folks were eager to take part in my research, but what if there was nothing for them to do? On top of all this, we had a film crew coming, and they, too, had great expectations of documenting our research and especially the elephants.

After we had organized all of our equipment, laid out the sound and seismic instruments, run test experiments with them, and then filmed the whole set-up, there were still no signs of elephants. We then took the time to tweak our electronic Observer data logger to further refine the behaviors and qualifiers we had programmed into the database, determined to be ready when the elephants finally showed up.

At last, in the late afternoon, after four days of waiting, our patience was rewarded by a visit from Greg and his right-hand bull, Torn Trunk, as well as a newcomer, Johannes. Greg's once mighty conga line had dwindled to just two followers. We were thrilled to see Greg and Torn Trunk again, but where was the rest of Greg's entourage? Where was his posse, over which he'd held such tight control? It was very unusual to see Greg with only two other members of his club.

Could it be that more choices of watering hole meant less pressure for forced social interactions at a single water hole and, perhaps, less of a need to kowtow to the don or seek even protection? And who was this new guy, who seemed so tight with our well-known Greg and Torn Trunk, and yet somehow we had never before seen him?

The three bulls had appeared out of the north like a line of three wise old men, walking slowly and purposefully toward the water with their ears flapping, first Torn Trunk, then Johannes, and finally, Greg. Torn Trunk arrived at the water hole first, looking a little uncertain about our operation on the other side of the clearing. This was fairly typical. The bulls were usually a little warier of us the first day or two after their arrival, but then after that they almost never heeded our presence. Torn Trunk gave us a sideways glance, then an open-mouth threat and slight trunk toss before settling in to drink at the head of the trough.

Torn Trunk was well ahead of the others, and as soon as Johannes entered the clearing, Torn Trunk stopped drinking, looked at Johannes and stepped away from the trough and positioned his rear end toward the incoming bulls. I wasn't sure if Torn Trunk was doing this as an act of supplication because Greg was on the heels of Johannes, or if it was because Torn Trunk was less dominant than this new bull.

Johannes passed Torn Trunk and marched straight to the prime position at the trough. This lasted for only a second, as Greg was right behind him and quickly shoved him out of the way. Once this score was settled, Torn Trunk drank next to Greg in second position, leaving Johannes downstream from them intermittently drinking and sucking his trunk, with Torn Trunk periodically throwing him threatening looks over his shoulder. This arrangement continued throughout their long drinking bout.

For our part, we settled in to score their behaviors, although with no younger bulls in the mix, we didn't expect much interaction. Still, we didn't want to miss key displacement events. We scored such behaviors as drinking, standing, orienting west, orienting east, heading east, heading west, looking, freezing, freezing with trunk on ground, orienting south.

Greg seemed to be dealing with the new social situation with great difficulty. Having been accustomed to his position as don, he always chose the departure direction and led the group out. But today things turned out a little differently. On this day, Johannes chose a direction and initiated departure. Greg cracked his ears in objection as Johannes ventured out south.

Greg assessed the situation and, oriented toward the west, placed his trunk on the ground and froze in position for a long time, as if to ascertain which direction was safest. Finally, he made the decision to head north-west, in the opposite direction from Johannes, with Torn Trunk dutifully following him.

Johannes held his ground and continued south. The situation became increasingly comical, as Torn Trunk seemed conflicted about whom to follow, and Greg more and more agitated by Johannes's insubordination. Johannes clearly didn't want to follow Greg's lead, and Torn Trunk seemed to vacillate as to whether he should follow his longtime headman or this new guy with whom he seemed already well bonded. He'd take a few steps to follow Greg, then turn and watch Johannes leaving in another direction and start to follow him and then stop and turn to follow Greg. Inevitably, Torn Trunk continued to follow Greg. This left Johannes by himself, and he eventually made a long arc around the clearing to tail the other two out.

Despite the bonded boys' club and their tight associations, some bull elephant relationships appeared to be fluid, with individuals and groups coming together and then going their separate ways, joining up and breaking apart for reasons yet to be identified. That shouldn't have been too surprising, since family groups also operated within fission-fusion societies.

What was surprising were the consistent patterns of association held between large numbers of bulls in the dry years. The fact that we had never seen Johannes before was most likely just an indication that the population was more extensive than we had yet documented, with relationships far more complex than we had been able to measure to date.

After this first sighting, the appearance of these three characters became a regular feature of the season. We were beginning to wonder if all seventy of the bulls we had been watching the previous year simply left the park, never to return. I knew that this probably wasn't the case, but I couldn't understand where all the other bulls had gone.

Over the course of the next few weeks, occasionally a few youngsters would attend the wise men. But all the while, there were palpable signs of insecurity exhibited by the don: the exaggerated, seemingly self-conscious acts of self-touching, combined with heightened vigilance—such as the constant freezing, scanning, and assessing alternative departure directions with trunk on the ground. The vigilant behaviors we had witnessed in Mike last season were now consistent features of Greg's behavioral repertoire.

To make matters worse, the altered environmental conditions supported more musth bulls, most likely because more females were coming into estrus. Because of the increase in musth bulls, it seemed increasingly difficult for Greg to maintain order.

Adding to the dynamic was that females had less tolerance for the antics of young bulls within the family, preoccupied as they were with tending to their larger-than-normal nursery. And therefore a larger than usual contingent of adolescent males had been expelled from their families. These young bulls contributed to the general uncertainty by being uncertain of themselves and seemingly fickle about where their loyalties lay as they negotiated new relationships within the boys' club.

But where, I wondered, were all the adult bulls? What had looked more and more like leadership behavior might merely have been Greg's acting for his own benefit. Otherwise, at least some of the other adults would have wanted to be with him, regardless of water availability. But what motivation would there be for Greg to lead, beyond simply sitting at the top of the totem pole to secure the best resources?

It's important to note the distinction between leadership and being the highest-ranking individual. Throwing one's weight around in order to secure access to the best food, drink, and mates, can lead to being at the top of the hierarchy, but such ranking is not necessarily indicative of leadership. Leadership implies making choices that would ultimately benefit the group, and thus, to follow such a leader would be advantageous.

For the female elephant, this could mean following an older matriarch to a remote water source during a drought, for example. There is a clear benefit to individuals for following this elder, and in the long run, there is an evolutionary benefit as well, given that the followers are more likely to survive than are females from other families with younger, and perhaps less knowledgeable, matriarchs. Consequently, followers of the experienced matriarch are more likely to pass on their genes to the next generation. The advantage to the elder is that her familial genes are being passed on as well.

For Greg's posse, it was still unclear that Greg was adopting a leadership position in his rallying of the troops and unclear as well what the benefit might be of following him, other than the advantage of being social rather than solitary—an advantage tempered by the group's need to share resources. Was there an evolutionary advantage for male elephants to form groups? And would a male elephant gain evolutionary fitness by leading a group?

These questions were worth pursuing, but without data on paternity, it would be hard to tell whether individuals within this male group were more successful at passing along their genes to the next generation than more solitary males, or that the leader had improved fitness by making choices for the group that benefited the group. One way this would work for males is if they were related in some way. I was going to have to do some genetic analysis on the fecal DNA we had collected. In the meantime, it was hard to ignore the shifting dynamic of the boys' club. It seemed like a kind of domino effect, where, one after another, the bulls in the boys' club had fallen away from the group.

Capo di Tutti Capi

▼▲▼▲▼▲▼▲▼

Just when we were wondering how long a dominant bull could hold on to his reign, Greg was displaced with no contest after two years at the top. It was late morning when I saw four bulls appear out of the west, just south of the tower. As they got closer, I could see that it was indeed the wise men plus an additional member of the boys' club, Keith Richardson, with his signature long and thick but scraggly tail and a C-shaped nick out of the middle of his left ear. Johannes walked in first, then Keith, then Torn Trunk, and bringing up the rear was Greg.

They walked slowly in a line, each glancing at us, up in the tower, from the side and sniffing toward us with the bottoms of their trunks. They seemed particularly wary of us and finally stopped after rounding the tower, all holding their ears out toward us, Greg sniffing in our direction with an upturned trunk.

They froze momentarily in a tight cluster and then continued on around the tower and toward the water hole, Greg remaining closest to us. They stopped again and assessed the situation. Greg gave us a foot toss and then started sucking his trunk as the others stood looking at us with their trunks on the ground. This was not the confident group of guys I had grown accustomed to watching.

Finally Greg flapped his ears and advanced, spurring the others forward as well. But suddenly he stopped and scanned with his ears again before moving on, dropping his penis as he walked. The others walked in a tight group a few paces behind their don. Greg stopped one more time and held his head up and ears out toward the tower, again sniffing us with trunk stretched way overhead. He then picked up the pace and walked up the southeast path, eyeing us all the while, as the others followed behind him.

They arrived at the edge of the pan and started to drink. Greg kept himself closest to us and continued to eye us from the side as he drank. He had never seemed this concerned about our presence in the past. When he appeared satisfied that the situation was safe, he commenced a mud bath with the others as they traversed the pan and reached their final destination—the trough.

At this stage Johannes displaced Torn Trunk from the head position, while Greg remained at the lower end of the trough with Keith. They still seemed wary of us, and at one point it seemed as if a zebra alarm call had startled them, since we'd made no particular noise to cause them to jump as they did. But it became clear over the course of the next few minutes that they viewed us as the cause of their alarm.

In the shuffle that followed, Torn Trunk mistook Greg's headshake toward the tower as an aggressive act and jumped out of the way, allowing Keith access to the head of the trough. After the reshuffle, all four stood frozen for about thirty seconds, Greg again giving the tower a watchful eye as he sucked his trunk.

At last, the four relaxed once more and went back to drinking. Then suddenly Keith perked up his ears and looked out to the west. A moment later, he held his ears out toward the tower and then they all drank again.

It seemed unusual that Greg would be this wary of the tower since he had already been to the water hole for a visit a few days earlier. Was something else going on out there to make him more on edge than usual?

Over time, Greg became increasingly agitated. He kept placing his trunk on the ground and then scanning to the west. Torn Trunk followed suit, and eventually the others did as well, the four bulls looking like the four points of a compass, each positioned forty-five degrees from the other. And finally, they all oriented to the northwest.

From the edge of the northwest clearing, in pranced a very musthy newcomer with two young companions. This new bull—whom we later named Smokey—gave away his true state as he marched in, waving his ears one after another and curling his trunk over his head like an enraged man waving a stick at intruders. He also was dribbling urine at an alarming rate. All indications were that Smokey was at the height of musth.

Greg immediately headed south and turned his rear toward Smokey, as did Torn Trunk who was heading east, while Johannes did the same in the southeast corner. They all watched over their shoulders as the two younger bulls arrived at the trough and started drinking. Keith was the only one that hadn't positioned his rear toward the incoming bulls.

Unperturbed about what was happening, Keith faced the musthy Smokey directly as Smokey approached him. It was clear, however, that Smokey was more interested in Greg, who was keeping his eye on him over his shoulder.

Smokey drank at the head of the trough, while the three wise men remained with their rears toward him, as if waiting for his next move. Keith went to join the two youngsters at the lower end of the trough, perhaps sensing that trouble was brewing among the elders.

Eventually, Torn Trunk slowly skirted the pan, making his way toward his buddies so that they could all head out together. Greg turned, half-facing Smokey in a retreat toward Johannes, and the three tried to make a quick exit.

Smokey, however, was right on their tails, dragging his trunk on the ground, dribbling urine and waving his ears in a musthy threat. We had never seen bulls give another male such a wide berth before, but Smokey had Johannes and Torn Trunk retreating to a distance of a hundred meters. And Greg, much to our amazement, was equally wary and dropped his penis in a full-bore butt-jiggling retreat.

Who was this new guy, Smokey? If anyone, we would have expected Kevin to be the one to overthrow the don. Given their extreme reaction to this bull, I wondered whether Smokey commanded such power outside of musth. At the moment, it appeared that, while Greg may have been the don of Mushara, Smokey was Namutoni's *capo di tutti capi*: the godfather.

Interestingly enough, Smokey knew who was in charge at Mushara and seemed just as agitated by Greg as Greg was by him. At one point, he was so agitated that he walked over to the place where Greg had previously defecated and performed a dramatic musth display over the offending pile of feces, dribbling urine and curling his trunk over his head, waving his ears and prancing with his front legs in the air, mouth wide open.

If anyone unfamiliar with elephant behavior were watching this display, it would have seemed like an expression of jubilation. But nothing was further from the truth. This guy was a testosterone bomb about to explode.

Greg is the first to veer off course when he sees Smokey at the water hole. Being in musth, we assumed that Smokey would be dominant, but this was the first time we witnessed Greg back down to any bull, regardless of hormonal status.

Greg finally left the clearing with Torn Trunk and Johannes at his heels, Smokey waving his trunk at them in fury. I had never seen such a dramatic retreat, and by no less a figure than the don himself.

Seemingly satisfied with the effect he had had on these bulls, Smokey returned to the trough to drink. This prompted Greg to return to the pan, apparently not having finished his drinking session. Torn Trunk and Johannes were slow to join him, but eventually followed suit, all now with penises dropped in supplication.

Smokey took one look at the bulls as they approached the water again and snapped. He marched over to the pan with shoulders and head high, dramatically scattering a herd of zebra in the process, and again dispersed the wise men, all the while dragging his trunk, waving his ears, and curling his trunk over his head to great dramatic effect.

The wise men watched the display from over their shoulders as they skirted the clearing south and then approached the water hole again. I assumed that they must have been really thirsty to keep this up, unless they were also trying to hold their ground by not leaving the area entirely.

Again Smokey pranced his front feet at their defiance, curling his trunk and swinging it at them, keeping up these displays until the others re-luctantly decided to head out, the younger bulls that had been trailing Smokey seemingly confused and scattered in the chaos. Smokey hung his trunk over his tusk as he watched the others retreat once again.

Keith, however, decided to remain behind. He seemed intrigued with Smokey and approached him while he drank at the head of the trough. I wondered what Greg was thinking as he stood at the end of the clearing, once again debating departure directions with Johannes, minus his loyal little apprentice, Keith.

Younger bulls seemed particularly curious about musth and, if permit-ted, they would inspect all of the physical attributes of a musth bull. Keith spent a lot of time visiting with Smokey, supplicating with a trunk-to-mouth greeting and then, sure enough, Smokey allowed Keith to inspect his very erect and green penis. (Because of all the urine dribbling, the penis on a musth bull accumulates algae, making it look disturbingly green in color.)

While his penis was groped by this younger bull, Smokey stood with trunk hanging on his tusk, allowing Keith's exploration of his testosterone-enhanced vigor. He then turned and initiated a gentle spar, after which they engaged in a mutual trunk-to-mouth greeting exchange. It was as if Smokey was sizing this guy up as a suitable apprentice. But since musth bulls aren't supposed to be interested in soliciting followers, this behavior was a little confusing.

Meanwhile, Greg had now turned around and was watching this dis-play, still appearing to be waiting for Keith to catch up. Even more amaz-ingly, he actually came back in and approached the male love fest from the side, causing Keith to desist from his display of reverence toward this potential new mentor.

Keith turned the other way, as if he were pretending that nothing had happened, as Greg bravely approached the low-ranking end of the trough. I was surprised that Smokey put up with this but also heartened that Keith actually approached Greg at that point, dropping his penis in supplication.

It only took a second for Smokey's patience to dissolve, however, and he scattered the two of them away from the water. At this point, Keith seemed to side with the more dominant of the two and remained with Smokey. Had Smokey just successfully stolen one of Greg's loyal boys' club members?

Not ready yet to surrender, Greg edged around the pan and took a defiant mud bath with Smokey staring at him from the trough. Keith stood between the two great titans with his trunk on the ground.

With Smokey's back to them, Torn Trunk and Johannes took the opportunity to head back in as well. As they were negotiating a safe distance from Smokey, Greg decided to leave, and this time on his own terms.

He headed up the northwest path, dusting aggressively and looking over his shoulder to see who would follow. Torn Trunk was the first, as expected. Johannes stared at them with his trunk hanging over his tusk, reluctant to follow, but eventually did, though not without balking a few times about the ultimate direction. The big question was whether Keith would also join them.

Keith watched the others leave, still with his trunk on the ground. Was he going to move up to the big-time players or stay loyal to the local don? He eventually chose the latter and headed out behind the others, while Smokey took off to the east alone. Sure enough, as a musth bull doesn't tend to appreciate hangers-on in his quest to his in rank, Smokey had abandoned his underlings.

With Smokey's arrival on the scene, there was now a bull that outranked Greg—a king with power greater than that of our Mushara don. Could Smokey have caused Greg to lose his confidence? Greg certainly acted as if he'd lost his edge.

However, the dethroning of this previously unchallenged bull could be temporary and part of the "turn-taking" found in previous musth studies— the musth bull trumps all. Kevin, of course, was our first exception to the rule, where musth did not trump the dominant bull. And that exception was worth trying to figure out.

At that juncture, we had yet to see Kevin and Smokey together. I suspected that Kevin was smart enough to avoid close proximity to this formidable bull who had just supplanted Greg. For the time being, Greg was still able to keep Kevin in line, so Smokey would have been completely out of Kevin's league.

After the incident with Smokey, the number of adult males visiting Mushara picked up, but there were additional signs of the erosion of Greg's power. It began with his gradual loss of command over the group's departures. In previous seasons, it was rather dramatic to see Greg lead his group out in a long line of twelve to fifteen bulls, initiated by stepping away from the water, fanning his ears and emitting a let's-go rumble, just as a matriarch of a family group would do. This was a simple but sure sign of the command that he wielded over his subordinates. The only challenge might have been a reluctant youngster that he had to give a shoving start.

Greg's ability to rally followers progressively diminished by the end of the season. We witnessed this one day when Greg arrived with Torn Trunk and Johannes as usual, but this time Tim, Willie Nelson, and Kevin were also in tow.

Throughout the visit, Greg seemed more preoccupied with vigilance than with socializing. But after the group had been at the water hole for some time, and Greg had made a decision to leave and head south, no one wanted to follow him. This was in striking contrast to the years before, where, every time he made the call—the low frequency let's-go rumble—it would be answered by a series of coordinated, nonoverlapping similar calls as the others lined up behind him. He'd then lead them out in one long line.

This time, when Greg made the call, no one responded, and no one followed. They just kept on drinking. Greg stepped away from the water, flapping his ears and giving the let's-go rumble again, but no one paid any attention.

It was like a scene out of a bad high school–themed movie: the boys are all out drinking in the parking lot, and the one who used to be the cool kid, but who's been getting upstaged lately by the newly arrived badass, says, "I'm outta here. C'mon guys." But no one moves, they just stare awkwardly at their beers. Which is just what Greg's friends did: they kept on drinking. Tim looked tentatively up at Greg, and for an instant it appeared as if he would follow—but he didn't.

It was painful to watch Greg as he stood in the middle of the clearing, waiting, hoping not to have to return to the water with his tail between his legs. He rumbled again. No response. He continued to move away from the water hole, then stopped at the edge of the clearing and again rumbled, but to deaf ears.

Again, Tim showed signs of being torn as to whether to follow. When Greg called, all the others kept to the business of drinking except Tim. He'd stop drinking, lift his head up, look at Greg, then look at Torn Trunk, then at Willie Nelson, and finally at Kevin, but could see that there was no impetus among the others to leave. He hesitated, touched his trunk to his mouth, as if still uncertain whether to follow Greg or hang with the boys.

Why was he uncertain? Was it some sense of loyalty that Tim felt for Greg? There was no way to tell at this point.

Greg stood at the edge of the clearing, faced with an embarrassing dilemma: should he leave the area on his own? Or should he skulk back to the water hole to join the others?—a downright humiliating option. He kept glancing over his shoulder to see if the others were going to make a move.

After several more unreturned calls to leave, Greg decided to turn around and head back to the water hole. I wondered how he'd handle himself when he got there. Would he beat on the others for not obeying his command? Would he start shoving the younger ones, physically forcing them to leave with him? Would he pretend that he really didn't want to leave in the first place? How was he going to save face?

Sure enough, he appealed to his most loyal underling, Tim. I understood why he wouldn't have gone to Torn Trunk first. Torn Trunk, although Greg's right-hand man, was not one to respond to affection from a contemporary. Greg was smart to go for a bull that was at least ten years his junior. Affiliative behaviors were more the norm when there was a distinct difference in ages. Besides, Tim was the only one who even deigned to acknowledge Greg's departure overtures.

Although it was clear that Greg knew whom he should approach to get results, it seemed that he went overboard in demonstrating affection toward Kevin, which was completely out of character—rubbing up next to him and giving him an ear-over-rear body rub. Nonetheless, after some gentle sparring with Tim, Greg gave a gentle touch to Willie and Torn Trunk before attempting his departure routine a second time.

Greg lined up again and gave the command "let's go"! This time Tim followed him out and, reluctantly, Torn Trunk followed behind.

Willie eventually followed suit—perhaps thinking it was better being with Greg and the gang than it was to be alone with Kevin. Willie had a lot of experience with Kevin's relentless sparring, which required more energy than Willie appeared to have.

But just as Willie was halfway to the clearing, with Greg and Tim at the tree line, Kevin caught up with him and started what turned out to be one of the longest sparring matches we had witnessed to date. It continued until well past sunset. And with the moon almost full, we were able to watch the contest without using our night-vision equipment.

The sparring went back and forth between the trees and the open scrub. Willie retreated, then held his ground, retreated, then held his ground. All the while the persistent bully kept up the challenge. Willy contracted his trunk and held his muscles stiff as he received shoves from Kevin's lanky trunk. Kevin was so relaxed and confident that he tossed his trunk around like a fire hose.

Both bulls kept their positions face-to-face and squarely even, so as not to risk the jab of a tusk to the flank. Dust rose in the setting sun, illuminating the two bulls in a furious orange, Greg and Tim tiny silhouettes in the far distance.

Kevin would lift his mighty head and throw his trunk at Willie. Willie held his ground, pushing back no more than necessary, as if he didn't want to encourage any further aggression, hoping to ride out the bully's adrenalin and call it a day. Willie's endurance turned out to be more than Kevin had probably bargained for, and the sparring continued into the waxing moonlight before Kevin decided to call it quits. And eventually, they both headed out warily in the same direction, Greg's attempts to impose discipline on the departure having ended in disarray and now apparently forgotten.

As Greg's influence appeared to have waned, I realized that the role musth played in bull society was probably even more complicated than I had thought. If the state of musth hadn't allowed the lower-ranking bulls within Greg's club to rise to the top of the hierarchy, would that still be the case when he didn't have control over the group? Was the potential for hormone suppression only possible in the dry years, when Greg was able to maintain his posse?

We were hoping the existence of a bull like Smokey might provide us with some answers. Was Smokey part of the background noise that Greg was forced to give a musthy "pardon" to, stepping aside with regard to competing for mates while still maintaining some control over the boys' club—despite the social chaos caused by this inordinately wet year? Or was Greg now just another washed-up don?

Of Musth and Other Demons

—————— ▼▲▼▲▼▲▼▲▼ ——————

Luke came into the clearing extremely agitated one day. He kept flexing his penis, which normally happens when a bull is particularly irritated with another bull or excited by the presence of a family group. But in this case, Luke didn't seem bothered by either of his buddies, Keith to the left and Jack to the right.

I quickly learned that there was another outcome of penis flexing— ejaculation. I knew there were other species out there that engaged in such activity, such as the epauletted fruit bat, which can lick its inordinately long phallus while hanging upside down, until the obvious outcome is achieved. This repetitive behavior can make their presence in children's zoos somewhat awkward.

Is it the demon testosterone that drives heightened sexual arousal? Doctors have been known to prescribe testosterone to increase sex drive, and yet, one study suggests that the state of musth is spurred on by sexual activity. That would mean that testosterone were the effect rather than the cause.

As previously mentioned, testosterone is important for mediating male reproductive physiology in all vertebrates. It's responsible for the physiological changes that occur at sexual maturation, including the production of sperm and secondary sexual characteristics. Testosterone is correlated with behaviors associated with mating, such as aggression and social dominance, with testosterone levels fluctuating drastically during the year, especially in seasonally breeding species.

Testosterone is one in a family of related hormones known as andro-gens. Androgens are secreted predominantly from the testes or are de-rived from testosterone, all responsible for related behaviors. Since we have markers only for testosterone in our study, it is the only androgen we have focused on and assume to be an appropriate measure of aggression and musth in and of itself.

If testosterone was the cause of the epauletted fruit bat's fervor, and our brutish bull Luke's escapades with his prehensile phallus, then surely, the heightened testosterone state of musth would bring these behaviors to a peak. But that doesn't appear to be the case. What exactly does a bull experience when in the highly irritable and unpredictable state of musth?

Since I couldn't interview a musth bull, I explored what was known about testosterone therapy to see if experience among humans might shed light on this question. In my search, I came across several national public radio interviews, including one where a patient had undergone testoster-one therapy for sexual reassignment surgery contrasted with another who had undergone testosterone suppression therapy in the context of prostate cancer.

At one end of the testosterone therapy spectrum, a woman wishing to change her gender described her experience on receiving her first testos-terone injection as part of the process leading to gender-affirming surgery. The patient vividly described a sudden surge of male sexuality—a sense of feeling like a raging sex machine in desperate need of expressing carnal urges.

The patient described doing things a women would likely resent and weren't appropriate responses to seeing an attractive women, such as looking at her breasts as she walked by. And the patient tried desper-ately to resist the temptation to turn and look at the woman's behind, but couldn't resist.

The patient went on to describe an experience of being at an airport newspaper vendor next to a mildly attractive woman. The patient imag-ined her feminine side would have wanted to ask this woman to join her for coffee. The male side, however, fantasized about cornering this women and convincing her to share a bathroom stall to accommodate the instant and erotic urges that were boiling within. The patient was taken aback by the strength of these urges and the vividness of her imagination.

In order to determine which of these effects were the result of the testosterone injections and which were the testosterone in combination with a preexisting psychological state, it would be important to conduct a controlled experiment in which a group of women planning to undergo gender-affirming surgery would receive either testosterone or a placebo (saline). A second group—this one composed of women not interested in sexual reassignment—would receive either testosterone or a placebo. The experiences across groups could then be compared.

Since these data were unavailable, perhaps a case reported on in the *Washington Post* in 2006 can offer insights on the effects of testosterone. The case involved a sex offender who had castrated himself in his jail cell. Although this practice is highly controversial and its ultimate effectiveness is uncertain, the prisoner has stated that he "had not had any sexual urges or desires in over two years. My mind is finally free of the deviant sexual fantasies I used to have about young girls."

In the second radio interview mentioned above—of the patient who had ceased producing testosterone as part of his therapy for prostate cancer—the respondent indicated that a lack of testosterone had put him in a state of listlessness, that he felt completely unmotivated during this period. He explained that he didn't feel depressed, necessarily, but that he had completely lost his drive.

Could this lack of drive be comparable to what lower-ranking males in other species experience when subject to socially induced testosterone suppression? Could the wild and confusing ride of amped-up testosterone be comparable to the surge of testosterone in an elephant bull's musth experiences? With that thought in mind after witnessing Luke's penis-flexing activity, I shuddered to think what it would be like to watch him come into musth.

Instead, we witnessed an event with an extremely young cow in estrus followed by a group of very excited young males, seven phalluses twitching in anticipation. Her family was in a state of upheaval as these seven young bulls trailed the group, creating a riotous affair at the water hole. An explosion of splashing and screaming ensued, as everyone tried to drink and at the same time. It seemed to be the elephant equivalent of a rave.

The volume of roaring and rumbling escalated as the young female, looking no older than around nine, was mounted in the middle of the pan by one of the rapscallions. I wondered why the older females didn't push him off, as he wasn't that old, but perhaps they were too tired. Perhaps they had given up trying to keep the hoodlums at bay.

But then I had another look around and realized he was bigger than

I thought. Once a young bull has left the family, he's at least the size of the older females and much stronger. There was most likely nothing they could do but attempt to minimize the effect of the chaos on their babies.

After a few unsuccessful attempts at penetration, the bull was evaded as the little female took off at a run once again, heading toward the east. It was hard to see any evidence of the phenomenon of "female choice" being at play here (in which estrus females emit loud calls to attract musth bulls, whose presence would minimize the possibility of being mobbed by younger males), but perhaps it was only a matter of time before a musth bull would pick up her scent and calls and rescue her from the unwanted attention.

Considering the inordinate size difference between males and females, one would think that male-male competition drove the evolution of males toward larger and larger sizes to successfully compete for a mate. Examples of animals having evolved toward increased size include the elephant seal and rutting antelope, though there are many more. The idea of female choice is that, rather than the males slogging it out with size and brawn to win access to females among other rutting antelope (as is the case with male-male competition), the females would be doing the selecting, irrespective of any male-male bashing behind the scenes.

Such selective forces are at play in many bird species, which is why the males are often much more showy than the females. Resplendent males and drab females are a telltale sign in nature that the females are doing the choosing. Those ridiculously elaborate (and, might I add, impractical for traipsing around in the jungle) peacock feathers on males are not there to scare off the competition. And the famous song, dance, and feather displays of the birds of paradise have a very specific purpose.

The lengths to which males go to impress a female in a society where female choice is at play are truly remarkable. In the extreme, this is seen in the phenomenon of "runaway selection," in which ever more elaborate features in males get selected by females, resulting in the genes controlling those phenotypes getting fixed in the population to a point where those exaggerated features (horns, feathers, showy colors, fancy footwork, etc.) become deleterious to a species to the extent that it could be threatened with extinction.

While watching this young female's seemingly desperate retreat, I understood why females might prefer mating with a musth bull, as other studies had shown. Perhaps females are smart enough to choose a mate who will fend off all the competition, leaving her in peace with only a single bull.

But how does the musth strategy differ from rutting, where a male creates a harem of females that he defends with great vigor? Because elephants live in matriarchal societies, the corralling of females is not an option. Bulls appear and disappear in the lives of females, and following a dominant male is not in the females' genes. Thus, male elephants have to resort to infiltrating a family group and either trying to get away with a sneak copulation in situ or engaging a female to break away from her family and sticking with him for a several-day period while he guards her. The latter behavior, called "mate guarding," has the obvious benefit of preventing another male's sperm from siring his would-be offspring. It's no wonder there are long distance calls for both estrus and musth to help suitable suitors find each other.

We had indeed seen the musthy Smokey clear the scene of other bull competitors with a few waves of his ears and prancing of feet. Without a musth bull around, however, a young estrus female could be in for a torturous ride, as the young rabble-rousers didn't take no for an answer. This could amount to many hours of running and utter physical exhaustion with seven or eight young bulls in hot pursuit. And one didn't want to think about how things might turn out when she ran out of energy.

And, in the confusion, it wasn't just the young females that were vulnerable. On one occasion, a young newcomer with a larger-than-life personality, Rocky Balboa, took a shot at chasing the matriarch, Big Mamma. Big Mamma got separated from her family and Rocky chased her around the entire clearing before she could escape him and return to her family, waiting at the edge of the northwest clearing looking noticeably distraught.

As soon as she approached her family to reunite with them, a most compelling ritual ensued. The family started roaring and rumbling and flapping their ears, urinating and defecating as they encircled the matriarch who approached with trunk outstretched to greet her seemingly long-lost family after only minutes of separation. Clearly such situations are highly traumatic for the females.

The episode conjured up images of other unfortunate females in the animal kingdom. In the springtime, gangs of male mallard ducks practically drown a female in their attempts to mate. Male dolphins are apparently no more chivalrous. Elephant seals have been known to crush their beloved in the throes of passion and in their attempts to beat the competition. Or what about the male sea otter that bites the female on the nose so hard during copulation that he can kill her, and worse, sometimes continues to mate with his dead lover? I suddenly had a heightened appreciation for the musth bull and his single-mindedness.

We witnessed several more of these wild events, most often during the new moon, when conditions for viewing were least optimal through a grainy-green view finder. There were times when the young males were so agitated that they mounted each other as well, such was their zeal for sex, practice or real.

When a mating event actually took place, jubilant postcopulatory rumbles filled the night air. The most interesting of the low frequency rumbles were the monotonic (i.e., flat in pitch) rumbles of unusually long duration—up to forty seconds of continuous vocalizing in a series of overlapping calls—seemingly from many of the family group members.

From my earlier work on the let's-go rumble, I knew that, on average, one of these rumbles lasted about three seconds (a little longer in males), or maybe a bit longer, depending on the context, with a bout lasting up to nine seconds as part of a let's-go departure rumble sequence, with one or two others possibly joining in after the dominant individual initiates leaving. But this postcopulatory sequence was different in that it lasted at least four times as long and sometimes longer. If elephants wanted to get the word out that a mating event had just happened, this was surely a resounding proclamation.

The fact that this seemed to be a spontaneous group activity, almost instinctual in nature, made me wonder if there might be an impetus behind these calls other than announcing that another elephant birth might be on the horizon in another twenty-two months. Having done some work on vocalizations and hormone priming in the bird world, I was familiar with the importance of auditory (along with visual and olfactory) cues that serve to "prime" a female to ovulate. Could these postcopulatory rumbles serve a similar purpose? Not unlike our lioness who needed the repeated act of copulation—on the order of a thousand—to stimulate ovulation? After all, doves need a certain amount of cooing from their lovers before ovulating, and many colonially breeding birds need the stimulation of group displays and specific chorusing vocalizations in order to mate and lay eggs.

The poor success rate of zoos trying to impregnate elephants is well documented. Perhaps what has been missing are these ritualistic postcopulatory rumbles as part of a successful mating repertoire for ovulation and for the union of elephantine sperm and egg to be completed.

Meanwhile, as the season wore on, the exuberance of the promise of sex died down. The peak of estrus seemed to be over and the season was starting to slow.

As the camp expanded over the years, it was a struggle at times for everyone to be content with the confined quarters, the remoteness, and the silence, particularly toward the end of the season. But fortunately, many, including myself, found solace in recipe creation. As for the rest, their iPods allowed them to survive the unsettling silence.

I found that good food helped keep camp morale at its peak and I enjoyed building on the recipes of Mushara's five-star "café," as did a few other inspired individuals. Our menu had expanded to include a roasted butternut pasta, a slight variation on my pasta butternaise (made with diced butternut rather than ground beef), butternut soup, sweet-and-sour peri-peri garbanzos, Greek cabbage salad, and hummus salad for lunch. The recipes for Mushara café expanded in scope and sophistication, along with the research.

But it was tough to keep the energy up in the kitchen toward the end. Everyone was tired, and as for me, I tended to stay up too late, absorbing the last moments with a lone elephant bull and his quiet drinking as the waxing moon hit the horizon. Or that moment with the rhino staring at his grumpy reflection in the pan, twitching one ear forward and back, and then the other, as if he, too, were searching for some explanation of the cosmos, like an early astronomer contemplating the universe of stars reflected in the pan, using his ears as a sextant—as if all of his questions would be answered in the repositioning of an ear.

On the morning before we broke camp, I awoke to a blood orange glow that lit the pan in a predawn flame. Keenly aware that it was our final day of fieldwork, I stared at the hushed landscape, the red pan so still that it almost felt as if my breathing would disturb the silence. For the moment, I had the earth all to myself, just before the sun would turn everything over to the wind and the birds.

A lion roared in the distance, jolting me out of my predawn reverie. It was time to leave Mushara again.

The Emotional Elephant

▼▲▼▲▼▲▼▲▼

The people of India assert that the tongue of the elephant is
upside down, and if it were not for that, it would have spoken.

Muhammad Iban Musa Kamal ad-Din ad-Damiri,
Hayat al-Hayawan (ca. 1371)

After the 2006 season, I was preoccupied with Mike. Mike had arrived at
the start of that season with one of his magnificent tusks broken and appeared to be in a lonely funk for its remainder. He was constantly sucking
his trunk and exhibiting other signs of uncertainty and self-consciousness
by repeatedly touching his trunk to his tusks, temporal glands, mouth, and
chest during his visits to the water hole. I got progressively more interested
in this self-touching behavior as a possible indicator of social uncertainty.

The behavior reminded me of our trained elephant, Donna, at the Oakland Zoo who exhibited similar self-touching behaviors in the context of
experiments we were doing concerning the possible seismic detection of
low-frequency vocalizations. To test this theory, we had Donna stand on
a metal plate with a shaker attached underneath it in order to deliver vibrations through the plate.

When we reduced the level of vibrations delivered through the plate
during the experiments, there was a point at which Donna started to fail

her challenge more often, getting less than 70 percent of her exam correct when the vibrations were at a threshold of her ability to detect them. She held her head and shoulders low and she touched her trunk repeatedly to her mouth and chest—the same behaviors that Mike exhibited in situations of uncertainty—seeming to indicate that Donna was affected psychologically by this failure.

Admittedly, Mike was not one for confrontation, but it was nevertheless amazing how long his shaken confidence lasted. And during his last appearance of the season, one of the many musth bulls that year arrived and came at Mike with head up, heels pounding, and trunk dragging and swirling on the ground between his front feet. Mike buckled, running to a safe distance before standing and sucking his trunk for so long that I regretted not noting the length of time.

The behavioral manifestations of uncertainty I'd witnessed in the wild made me wonder whether our work with Donna and her force plate could help me better understand the impact of social pressures exhibited on the lives of elephants at Mushara. Could posture really tell us about an elephant's psychological state? Dare I ask if elephants indeed had what humans would consider to be emotion?

Until recently, emotion was a term referring only to human physiological, psychological, social, cognitive, and developmental states. The perspective on emotion is slowly changing, however, and there is mounting acceptance of the idea of using nonhuman animal models of emotion that might provide unique insights and a deeper understanding of emotions in humans.

Cognitive faculties such as perception, attention, learning, memory, and motor control are all influenced by emotion. Understanding the interaction between cognitive processing and emotion may facilitate better understanding of human medical disorders in which impairments of both emotion and cognition are evident, such as schizophrenia, depression, addiction, autism, and even Alzheimer's disease.

Since elephants share similar social behaviors with humans, they could be an ideal animal to understand human emotions from a new perspective. And I now had a mechanism to test whether I could simulate the emotion of uncertainty that I had witnessed in the wild (within known social contexts) under similar conditions in the lab.

Returning to Donna and our controlled laboratory setting at the zoo, we were, in that situation, in a position to test psychological uncertainty as it related to a particular mental challenge by simulating the conditions of uncertainty. When we presented Donna with a choice between two familiar outcomes that became harder and harder for her to differentiate, we witnessed indicators of emotional state, including changes in posture, such as hanging the head low, and gestures, such as self-touching of the temporal region, mouth, and chest with the trunk. These were the same behaviors that Mike exhibited in the context of social uncertainty.

Fortunately, we had already done most of the hard work in this area when we collected data on a vibrotactile threshold of Donna's feet. Donna had to make a yes or no decision and then touch the yes or no target with her trunk if she determined that the plate was indeed vibrating. The sweet spot for measuring uncertainty was when the signal got softer (by six decibels) and, thus, more difficult for her to determine whether she felt any vibration.

If we could learn about human emotion from elephants, I wondered what we could learn about early childhood development from these long-lived large social mammals. There were a lot of great studies on early childhood development conducted on nonhuman primates, where parallel intelligence tests had been given to monkeys or chimps (symbol searching, fitting round objects into round holes, color matching, etc.) and then the results were compared to those for humans. Elephants could provide a different and perhaps even more revealing model of early childhood development than primates, considering that the brain of an elephant calf is half the size of that of an adult elephant, whereas in humans, an infant has only 25 percent of the adult brain capacity, which means there'd be a lot more real estate for cognitive processing. An alternative model for understanding the ontogeny of emotion might provide insights into a child's understanding and experience of emotions, decision making, activity level, sociality, and confidence, as well as emotional communications occurring between parents and children.

As previously mentioned, elephants, like the great apes, are long-lived, highly socialized animals. They live in hierarchical groups within an extended network of relationships, making up a complex fission-fusion society in which breaking apart and reuniting is commonplace. Living in such extensive social networks is thought to correlate with and most likely promote cognitive sophistication. One of the hallmarks of this level of sophistication is the ability to modify tools (considered more demanding cognitively than the mere making of tools)—an ability elephants have been shown to possess. In one study, for example, elephants were shown to use their highly muscular prehensile trunks to modify branches for optimum use as switches to repel flies.

Although some caution that adapting a branch would be considered an extension of browsing behavior to tool modification and is not overwhelmingly impressive for an elephant when compared to the complexity of the ant- and termite-dipping tools made and modified by chimpanzees, or the serrated leaf probes and hook tools made by crows in New Caledonia—by definition, elephants have indeed engaged in tool modification. Perhaps more sophisticated efforts at tool modification remain to be quantified. For example, many anecdotes record, and I have also personally observed, elephants using creative tools to short-circuit an electric fence in order to access a crop or a water installation, including protecting themselves from getting shocked while breaking an electrified fence by using a tusk or even holding a rock to the forehead or throwing a tree into the fence.

In terms of cognitive processing, not only do elephants have the largest absolute brain size among land mammals, they also have the largest temporal lobe relative to body size of any animal, including humans. The temporal lobe is that portion of the cerebral cortex devoted to communication, language, spatial memory, and cognition. Given the temporal lobe's relative size in the elephant, there is every reason to suspect that elephants may be capable of far more complex cognition than is currently understood or documented.

In fact, elephant brains contain as many cortical neurons as human brains and have larger pyramidal neurons (specialized neurons thought to play a key role in cognitive functions) than do humans, suggesting that elephants might have learning and memory skills superior to ours. On top of this, von Economo neurons (or spindle cells)—believed to be involved in social awareness and the ability to make quick decisions and thought to exist only in humans, great apes, and four species of dolphin—were recently discovered in elephant brains.

How do these brainy animals take advantage of their great capacity to learn and adapt? Again, African elephant society is largely matriarchal. Dominant members of family groups are thought to make decisions for the herd as a whole about safety, movements, where to feed and when to approach the watering hole, and who to avoid and befriend. It makes sense that long-term memory would be a very important survival tool for this species. Scientists have documented (through satellite tracking and mapping) the fact that elephants use specific paths, migrating along them seasonally, and that these migratory pathways have existed for many years.

A 2009 study suggested that certain elephant family groups, compared with other family groups, experienced lower calf mortality during a severe drought in 1993 in Tanzania, perhaps due to the fact that older matriarchs within the extended families had experienced similarly severe drought conditions between 1958 and1961. The scientists proposed that these matriarchs may have remembered the earlier drought and led their families out of the park to areas where there was enough food and water to allow them and their offspring to survive. Can we attribute this increased survival (or reproductive fitness) to the long-term memory of older matriarchs? It's an intriguing possibility.

In experiments of memory over shorter time scales, elephants have been shown to distinguish calls from both known and unknown elephants, suggesting a memory of the calls of specific individuals. This memory cache would give older matriarchs the advantage of knowing whether a seeming threat was real, so that they could respond and steer the rest of the group appropriately.

The earliest tests to determine elephants' ability to discriminate and remember were done in the 1950s with a captive Asian elephant. A juvenile female was taught twenty different visual discrimination pairs and six acoustic discrimination pairs. It apparently took her 330 tries to learn the first visual discrimination pair challenge, but by the fourth challenge, it took her only ten trials to learn the correct answer. This study demonstrated that elephants can learn quickly. In addition, the experimenter repeated the study a year later, and the subject was able to remember the challenge and performed with 73–100 percent accuracy.

Studies suggest that elephants have the capacity to learn feature and category discriminations—an ability previously confirmed only in primates, pigeons, and rats. Elephants, however, appear to demonstrate classification abilities—the ability to group novel objects into sets by various traits or qualities—that might just surpass those of many other vertebrates.

A field study conducted in Amboseli National Park showed that elephants could discriminate and categorize objects that had ecological relevance. Researchers presented elephants with T-shirts and other items of clothing that gave either visual or olfactory information about two different human ethnic groups, the Maasai and the Kamba, that posed different levels of threat to elephants.

Traditionally, young Maasai men speared elephants in a rite of passage, while the pastoral Kamba people had no such tradition. In the first experiments, elephants were presented with T-shirts and other garments previously worn by either a Maasai man or a Kamba man. They showed significantly greater flight reaction to garments worn by Maasai, running away when presented with them. The elephants appeared to associate the smell of the Maasai with a threat, perhaps even associating the smell with the memory of an elephant being speared.

In a related experiment testing visual associations, elephants were shown either a white (neutral) garment or a red one (a color traditionally worn by Maasai). Elephants are dichromats—that is, having the same color discrimination ability as a color-blind human. White and red are distinguishable from each other, and elephants showed a significantly greater flight response to the red garment than to the white. In a more recent study, elephants were shown to recognize language differences between human groups and assess danger accordingly, even to the point of discriminating gender.

Considering the facts that this long-lived, highly intelligent mammal has a huge temporal lobe, highly sophisticated neural circuitry, and the largest brain capacity relative to any other mammal, the elephant is a natural focus of cognition experiments. Scientists have made progress on assessing elephants' visual, vocal, and olfactory discrimination, but other cognitive experimental questions are easier to pose than to investigate. Experimenting with elephants poses elephantine challenges.

Scientists rely on white mice, zebra fish, and fruit flies as study animals for a reason—they are cheap to raise and house. It is easy to create a controlled environment for such experiments and to run repeated trials to generate robust data sets. Comparative cognition work has been done on pigeons, pigs, dogs, and primates, but scale up to a study with elephants and it becomes much more difficult to find enough study subjects and run repeated trials.

Researchers have managed to conduct pioneering studies using zoo elephants in the areas of hearing, vibrotactile sensitivity, and self-awareness. Yet these three individual studies were limited to a sample size of one. (The self-awareness test was performed with three elephants but only one produced usable data.)

Scientists have long considered the ability to recognize oneself in a mirror to be an index of high cognitive ability and one that is associated with humans, apes, and other highly social animals. To pass the mirror test, an animal has to respond to its own reflection in ways that make clear it sees itself in the mirror, as opposed to thinking it sees another animal of the same species. In the classic test, the experimenter surreptitiously applies a mark or sticker to the study subject, then presents the animal with a mirror. If on seeing its reflection the animal looks for the sticker or mark on its own body, it passes the test.

Two such experiments were done on Asian elephants to determine whether visible or both visible and concealed markings would be explored by elephants in front of a mirror. Neither of the two elephants in the first study reacted to their reflections. In the second study, one out of three subjects explored visible markings on her forehead, an indication she knew she was looking at a reflection of herself and not at another elephant. Although not as indisputable as the responses of great apes, the results were significant enough to warrant further investigation.

Perhaps future experiments affording the opportunity for elephants to explore the mirror outside of circumscribed testing times, such as incorporating mirrors into elephant enclosures, would allow more individuals to respond, thereby leading to a stronger result. Such modifications may well demonstrate more definitively that elephants have a concept of self. In the meantime, these same researchers have shown the elephant's ability to empathize with the misfortune of another and console the other after a traumatizing event (similar to what we often see with an older sister or aunt when a baby gets stuck in the mud or is accidently separated from the group).

I couldn't help being tempted to explore Donna's cognitive ability by swapping the vibrational cues we had used for our vibrotactile studies for memory tests using pictures and objects. This time, I brought in some colleagues who knew more about this field than I did: Francis Steen and Dwight Reed from the University of California, Los Angeles. Together with the staff at the Oakland Zoo, we hoped to assess elephant cognition in two phases. First, we needed to establish whether elephants could recognize a picture of an object (say, a banana) as a representation of an actual banana. Next, we needed to show whether elephants presented with a picture of that banana could use the symbol to accurately predict the location of an actual food reward—in this case, a real live banana. Retrieving

Donna, a captive elephant at the Oakland Zoo, uses her trunk to recognize a picture of a banana (among three image choices), after being presented with the real thing.

the food reward would require Donna to both recognize that the picture of a banana stood for a real banana and think ahead and make a plan in order to retrieve the fruit. Could she do it?

A central feature of higher-level cognitive ability to plan is the ability to stay focused on a goal or object when there are no reinforcing cues. For most animals, once an object is out of sight (or smell, or hearing) it is literally out of mind. Our own capacity to generate mental images of things we remember seeing may have evolved in response to continuing selective pressure. The ability to formulate, test, and carry out individual plans of action may have given our species an advantage.

When designing tests of cognitive ability, it's important to differentiate between an ingrained magnetic or olfactory sense, such as what insects, turtles and birds might use to navigate (path integration), and an actual cognitive use of memory, with referential images of migratory paths that passed a favorite fruit tree or seasonal water pan. The ability to treat a representation (a picture of a banana, for example) as if it were the real thing (an actual banana) is required to form a mental simulation, in which inferences are generated from a memory stimulated by the picture.

As I wheeled that first wheelbarrow of bananas toward Donna's enclosure, I knew the process of investigating the cognitive capacity of elephants would be a long one. Testing their ability to recognize representations of objects would be an important first step in showing that elephants might be candidates for the kind of higher thinking called cognition. Showing recognition of objects alone wouldn't make a strong case. But before I could plan out the whole series of experiments, I needed to see if this first one would work. And if so, I hoped to be in a position to pursue the question of elephant cognition with Donna, one wheelbarrow of bananas at a time, wherever it might take us.

Once Donna had acquired the skill of touching targets with her trunk in response to an experimental stimulus, we began the study with our first challenge. We attached photographs to the targets and trained Donna to touch the relevant photo on the left or right target when a piece of fruit, either a banana or an apple was presented in front of her. She was allowed to eat the fruit if she selected the correct photograph.

Many, many bananas later, we began to realize Donna, at least, was not the most enthusiastic study subject. This was unexpected because she had become an expert on touching targets in response to vibrotactile cues and actually appeared to enjoy the challenge. But when asked to make decisions about a photograph and engaging in visual discrimination, Donna was clearly not paying much attention at all, although she was clearly enjoying her banana windfall.

Donna's trainer, Colleen, had to wave her hand in front of Donna's eyes constantly and snap her fingers to get her to pay attention and actually consider the photos. Donna would lift her heavy eyelids for a few moments, look up at the photos, and then plunk her banana-smeared trunk tip indiscriminately onto an image, breathing out heavily. Banana breath and banana slobber showered the photos as well as ourselves.

We could tell that Donna was trying to discover the rules of the new game but was having trouble figuring out what we wanted her to do—to make a visual choice, either between a banana and a blank white image or between an image of a banana or an apple.

Every once in a while, we'd test Donna's focus by swapping out her banana treat for an apple when she got an answer correct. Now *that* got her attention. She immediately opened her dozy eyes wide and looked around confused as to why the protocol of her getting fed a continuous stream of bananas had suddenly been switched to something else.

Our testing did not progress nearly as quickly as we had hoped and proceeded far less quickly than had the vibrotactile studies. Part of the problem may have been that elephants don't use vision as their primary sensory input. They tend to smell or hear something first, and only then train their sight on it. So we had a difficult time getting Donna to focus on our lovely photographs of plump and juicy fruit.

That vision was not Donna's primary line of sensory information about her world was borne out by the results of our previous experiments. When we tickled the bottom of Donna's feet, her focus was entirely on the vibrotactile stimulus. The fact that the visual sense is not dominant in elephants could explain why an elephant in the wild might become startled by the unexpected proximity of a human, even perhaps fatally trampling the unfortunate person. If a human is downwind and silent, an elephant's preferred senses of smell and hearing don't provide the early warning they might need to adapt to a person suddenly appearing in their field of vision.

Eventually, though, Donna got the hang of this new challenge and was finally succeeding at selecting the correct photograph more often than would be due to chance. We decided she was ready to be trained for the next phase of the experiment.

In the second phase, the goal was to test Donna's ability to plan by presenting her with a photograph of a hidden item, such as a pumpkin (a favorite treat). The photograph would incorporate cues of its location that were familiar to Donna, such as the fifty-gallon drum that dangled in the bull yard or a favorite tree stump out in the exhibit enclosure. If Donna responded to the image of the pumpkin by searching for it, this would be evidence that she might be capable of using mental imagery to formulate plans.

The critical factor here would be if Donna's search were initiated (and perhaps sustained) based on her mental image of a pumpkin rather than on visual or olfactory cues. That is, we would need to make sure to distinguish between, for example, a dog's act of searching for the source of a smell and an elephant searching for a pumpkin based entirely on its depiction and not on being smelled or on a sighting of the actual object. And to rule out olfactory cues altogether, we'd hide a picture of the pumpkin, and only present a real pumpkin as a treat for finding the photograph. If Donna was truly able to evoke the image of the pumpkin on her own without any prompting either by photograph or by smell, then cognitive scientists out there would be satisfied that Donna had indeed engaged in cognitive thinking. That's a lot of work, but it's the only way to move from anecdote to experimental proof, even though anyone that knows elephants in the wild can imagine them moving north after the rains, heading on a path that they always take, knowing exactly where the marula trees are, and when the fruit would be ripe, all of which seems to demonstrate cognitive thinking.

As Donna sucked on her umpteenth banana, I contemplated a debate among cognitive psychologists over the definition of cognition in general and then, specifically, in relation to whether bees were capable of cognitive processing, as, for example, in the interpretation of the famous waggle dance (a figure-eight dance used as a form of communication by honeybees). This argument—played out in the pages of various scientific journals over the past decade—caught my attention simply because of my quandary over Donna's struggles with her seemingly simple task.

The cognition community was turned upside down by the question of whether individual bees, in all of their mental processing ability, could be considered to have cognition. The flurry of discussion came when skepticism was voiced that creatures without backbones were capable of possessing this capacity to any degree. And at the same time, there were cognitive scientists who supported the idea that robots had cognitive ability. The field obviously needed a definition of cognition that could apply as well to people and primates and dolphins as it could to robots and bees and elephants.

Hence the debate continued over an official definition of the term "cognition" and the concept of reference—whether the representation of a real object could interact internally with another representation of reality. In other words, whether viewing the image of an object could stimulate a mental image of the same object—that is, whether the picture of a banana would cause Donna to imagine a real banana. Since Donna connected the act of touching the image of a banana with a banana reward at a rate slightly higher than chance, this suggested she might actually be thinking and not just enjoying the taste of bananas.

But the results didn't give us confidence that we were going to get definitive answers quickly. We decided to continue our pursuit, albeit more slowly than we had initially planned, in the hopes of learning something that would help elephant scientists and conservationists better understand the mechanisms underlying elephant cognitive processing in general and, perhaps, better predict outcomes and assist in the management of wild herds, which were under pressure from human-elephant conflict, shrinking habitats, and poaching.

Meanwhile, back at Mushara, Mike sucked his trunk beneath the blood-filled moon. I was eager to watch the denouement of his dilemma, but was also OK with waiting and wondering. How did each bull choose his strategy, if in fact there was a strategy? Judging from the diversity of characters out there, it seemed as though individual shaping and choices were involved. How did an elephant bull choose his company, and why were some bulls only acquaintances and others almost if not entirely inseparable?

I knew that wherever there was an answer, there was also a new question lying within it—which is the true nature of scientific exploration. But I also knew that, even if there would never be a complete answer, further study *would* at least provide a deeper understanding of this most complex animal.

We had one final visit from Mike the evening before we had to break camp for the season. His gait was so light and slow that he seemed to drift cautiously into the clearing, avoiding even the double-banded sand grouse lining the edge of the pan—a ghost of his former self. He had been a great mentor and gentle giant, a combination of traits seemingly irresistible to the next generation in many social animals, including humans. I had the feeling that he would rise again with a band of loyal apprentices of his own. It was just a question of whether elephant-bull society would tolerate a benevolent softy as a dictator, and whether he was motivated enough to maintain a posse of his own. With that thought in mind, I turned my attention back to the uncertain political position that Greg was navigating.

The Don Back in the Driver's Seat

———— ▼▲▼▲▼▲▼▲▼ ————

On the first day of an extremely dry 2007 season, the camp was abuzz with our new digs: a three-story tower with twice the floor space on each floor, and placed on the northern side of the pan rather than the southern side. The first layer would be dedicated to tent space, which simplified the camp setup tremendously. Granted, the tower was still of the same basic form it had been when I left it at the end of the previous season—a simple frame with no walls and no floors—but the difficult tasks of welding the railway-line steel structure together and cementing it into the ground for extra elephant-proofing had been completed.

No sooner had we unloaded all of the trucks, including three floors-worth of wood, when the boys' club arrived in full force—seven males with Greg in the lead. They were undeterred by our disarray and headed straight for the water hole.

Despite all of the setup duties to attend to—we had only a few hours left of sunlight and needed to get the boma cloth up to secure the camp before nightfall—I couldn't resist taking a moment with the boys, particularly after such an atypical season in 2006 with only a few members of the club showing up at any one time. Finally the boys' club was back intact, and I was eager to get a sense of the mood of the group.

I quickly set up a table under the tower for my camera bag, and started taking pictures while the others continued the unpacking. Flanking Greg on his left were Torn Trunk, Prince Charles, and Dave, and on his right were Keith, the formidable Marlon Brando (fairly new on the scene), and Tyler. Now, *this* was going to be an interesting year.

I had purposely delayed our arrival by two weeks just to ensure that the elephants would have returned to the site. And sure enough, here they were, looking like the reservoir dogs of movie fame with Greg back at the helm. With such a dry year, it was like the good old days of 2005 with much larger groups. I assumed that a strict pecking order would also be in place.

Although Marlon Brando was going to be a serious challenge for Greg, I looked forward to seeing how that relationship would develop over the season. I also wondered whether Smokey would return, or if perhaps Mushara lay outside his usual range, and the inordinate rains of 2006 had sent him wandering farther afield than usual.

In addition to getting the boma cloth up first thing, we also needed to set tents up, which together would allow us to sleep safely on our first night. We then pressed on with the rest of the camp construction despite a fairly strong wind—digging fence postholes and a hole for the long-drop toilet, assembling the kitchen tent, the dung-analysis station, and sound equipment tents.

New floorboards were laid on each level of the tower, and ladders put in place between levels. After that, we put up personal tents, and slowly the camp was coming together while the boys' club looked on from the water hole. Tower reinforcement cross beams would be dealt with the following day.

We had a reporter from the *Baltimore Sun* with us during setup who documented our progress as well as the presence of the boys' club. There was a brief interruption when a pigmy python appeared at the base of the tower after a few fence poles were removed from their resting place. The python was placed in a cooler to safely contain it, driven to the perimeter of the clearing, and released. It was unusual to see snakes in the dry season, but pythons were much more welcome than the assortment of poisonous and aggressive snakes that inhabit the park and that we had been fortunate enough never to have seen at the site, most especially the black mamba.

After the floorboards were placed on the third floor, I climbed up to have a look. I was a little nervous about the height of the third level at eight meters and wanted to see if I would feel as I had on the last day of the previous season when I climbed to the top of the partially completed new tower and felt somewhat dizzy at the unexpected height.

The view was much more expansive than that from the old tower; you could see much farther beyond the trees and into the distance on all sides. Standing on solid floorboards made a big difference in dispelling my previous hint of vertigo. And I knew that once we welded cross beams on the sides and back of the tower on each floor, the slight swaying sensation would disappear.

The elephants looked that much closer from the eighty-meter distance of the new tower in contrast to the 120-meter span from the old tower. The subtleties of postures and expressions were going to be that much easier to document.

I could see that there was still tension between Prince Charles and Tim, and I was surprised to see them within the same group. Tim was keeping his distance. Tyler was going to be interesting to track after his previous testosterone swings. He was currently hanging tight with Marlon Brando, and this seemed a smart choice, as Torn Trunk had probably lost patience with him. Dave was up to his usual sucking-up-to-the-don routine.

Meanwhile, down below, the kitchen tent and tables were up and the kitchen was getting assembled. The perimeter poles had been set and the five strands of wire strung around them to support the boma cloth. The air had reached its late afternoon stillness, and the lack of wind meant that it was time to roll out the boma cloth that would then get attached to the wires.

Not wanting to miss out on any of the boys' club action but still needing to oversee the operations on the ground, I put a volunteer on video duty so I could watch the tape later and not overlook any significant behaviors. I set up the video camera, tripod, and a chair and gave a lucky volunteer a break from manual labor.

Climbing down the ladders between floors, I again compared the differences between the old and new towers. The main difference was the new location, across the water hole facing south, just as the ministry requested so that the view from the bunker would not have a tower in it. This new vantage would take some getting used to, but the subtleties—such as the tower's position in relation to sunrise and sunset—were only apparent later. The old, north-facing tower located in the south clearing provided much better vantage for experiencing the reds and oranges of these two daily events—an aesthetic sacrifice to be sure, but more than balanced by the enhanced conditions for observation.

I checked in on my Stanford student, Mindy, who had disappeared into the long-drop hole, having taken her job of digging the hole extremely seriously. She never relinquished the shovel and was determined to finish the task on her own. We all had a good laugh as she climbed out of her hole, took a breath, and wiped her brow. She had completed her mission and the long drop was ready for action.

The sun was setting just as we enclosed the last few feet of camp, and a lion roared in the distance—the pressure waves from this primal call penetrating our chest cavities and causing our hearts to flutter—a reminder that we have not conquered nature completely. We were still prey after all. And with the closing of the gate, the 2007 season had officially begun.

Throughout the season, Greg retained command of his club. He headed a long train of elephant bulls of varying sizes, and when they arrived at the water hole, the bulls allowed the don to drink unchallenged at the source of the spring. With water only available in a few places, the hierarchy was once again in place.

Interestingly, ranks hadn't changed much—even after the previous year's chaos. And once again, the boys arrived with Greg, and, yes, even left on his suggestion, initiated with a let's-go rumble. Everything was back to normal for the local club—with one notable change.

Greg was less of a tyrant this year than last. Perhaps he realized the power of persuasion was a more successful technique for remaining in power than was the use of punishment. Perhaps he knew that, in order to stay on top, he needed to give a little extra attention to his supporters—having seen the previous year what could happen when his rallying of the troops was ignored. Maybe he recognized that he needed to modify his strategy and be a little softer on his mates.

This didn't mean that Greg wasn't stubborn. And he did exhibit more aggressive behaviors overall than any of the other bulls over the course of three seasons. But, it was gratifying to see that he was able to adjust his behavior, toning down his aggressive interactions when things were under control with his posse.

This makes sense in relation to the theory of why animals form hierarchies. When a linear hierarchy is in place, less aggression is needed to keep the pecking order straight, since everyone knows where he stands, and thus conflict is minimized. And when there was no hierarchy in place—as exhibited in our second season—all of the bulls exhibited more aggressive behaviors. This may have been due to their uncertainty about each individual's overall rank in the pecking order.

Not everyone was happy with his ranking this season. Late one afternoon in July, while busy with a photo shoot of the boys from the ground-level bunker, I was treated to a close up look at Greg's aggressive posturing toward Captain Picard. For some reason, Greg had it in for the Captain. Every time he showed up, Greg became extremely agitated, and, without any hint of instigating others to follow his lead, Torn Trunk was there by his side to participate in the bullying.

This time, Greg gave the Captain the most exaggerated ear fold I had yet seen from him and then performed a huge stationary leap with his trunk fully extended. It looked like such a funny threat, yet he meant business. It was one of those Incredible-Hulk moments when Greg pumped up, flaring out his neck and clenching his teeth (which in elephant language was an open-mouth threat) at his opponent without any forward movement. Seeing that the Captain continued to approach the water, Greg marched over to confront him head on.

The Captain seemed genuinely surprised at the sight of Greg bristling toward him. Or perhaps Greg's posturing was the reason for the Captain's long delayed entry, but thirst eventually got the better of him.

Apparently, Torn Trunk was taking this all in, as he stopped drinking. He marched over and positioned himself next to Greg with his ears out and head up to back up Greg's threat, effectively blocking the Captain's entrance to the water. The Captain had placed himself in an extremely tenuous situation and now backed down. Shuffling along with shoulders rolled forward, he made his way all the way around the pan and sipped at the muddy water.

This all transpired against a backdrop of the usual cast of characters, who, on this occasion, steered clear of the display and carried on with their drink. There were Willie Nelson, Kevin, Tim, Keith Richards, Luke Skywalker, Jack Nicholson, Prince Charles, and the new Frankie Fredericks, along with another new young bull with perfect ears that I was able to photograph up close. It was always more difficult to characterize the very young bulls, as their ears had less wear and tear than the older ones. All the while keeping tabs on what was happening to the retired Captain, I was able to get good identification shots.

I don't know what he had done or why he was being ostracized from the boys' club, but for whatever reason, Captain Picard was definitely on the outs with everyone this season. He was even rejected by the low-ranking Tyler. He was one of the oldest bulls at our site and, as far as we knew him, he always seemed rather innocuous, dipping in and out of boys' club affairs without ruffling feathers but never really with a long-term presence within any one season.

This season was different. Captain Picard's insecure posture had become so obvious that we could pick him out far in the distance, standing within the immediate tree line, cowering for long periods with his fragile shoulders hanging low. It was as if he'd come to expect rejection even before entering the clearing and waited until the last thirsty minute to have to experience it. This didn't seem logical, however, as he could have timed his arrivals outside the boys' club visits. But for whatever reason, he was on a satellite course with the band of merry men, wanted or not.

His last visit was made particularly poignant by his treatment by a new very young bully named Malfoy. The Captain was quite innocently standing at the head of the trough drinking peacefully when the young menace sauntered in and sidled up across from him at the head position. Captain Picard reached out to give the new bull a trunk-to-mouth greeting and, in a sudden burst of violence, he got a head thrust and trunk slap instead. What flipped Malfoy's switch was a mystery.

Having no respect for his elder, Malfoy pursued the Captain with such intent that the Captain forgot the rest of his drink and ran for his life. I didn't think he had it in him to run so fast.

Malfoy chased him all the way across the clearing, the Captain running with his stiff legs, thin frame, and bobbing head. What could he possibly have done to deserve such treatment? He hardly could have been perceived as a threat.

But maybe that was just it. Perhaps the Captain's frailty was not a trait that the others wanted to be around. Could he have been ill and the others able to detect it? Is it possible that they might think he was contagious? Why else would he be rejected by both young and old?

Such behavior is common in the animal kingdom, including in humans. And since anthrax was natural to the environment of Etosha, and elephants died every year from the disease, perhaps these other elephants could see the writing on the wall for those about to succumb. It was indeed a mystery.

As I continued my photo shoot in the late afternoon sun, who of all elephants turned up but one looking very much like little Congo Connor. We hadn't seen him since 2005. I took the camera down from my eye so I could get a good look at him. I picked up my binoculars and looked at his ears. Notch out of middle of left ear, left tusk lower than right. Broken right tusk. Small rounded ears. It was Congo all right.

There he was, as if he had just been in the day before. He walked right into the scene with the other bulls and gave Tim and Willie a big hello. He placed his trunk in Tim's mouth, and Tim leaned into him as if to say, "Hey little guy, where have you been?"

Congo then said hello to Willie. First he placed his trunk in Willie's mouth and, after that, he held onto Willie's magnificent up-curved tusk with his trunk and leaned in as if giving him a slap on the back while shaking hands. He definitely had a flare to his greetings—like a surfer dude or a skateboarder, holding his trunk out to his buddies for a high five after coming off a great halfpipe. I was committed to getting to the bottom of his confident suave self.

With the setting of the sun, a family group arrived just as the bulls began leaving. This time Congo was heading out with his young mentor and buddy, Tim, who followed on the heels of an orderly boys' club departure (except for Willie who appeared to be in the mood for lady friends rather than the boys). I hoped that this was the beginning of a habit for Congo as I wanted to see more of this spirited young male.

As Congo and Tim left the water hole to the west, I took photos of Willie and his female admirers before it got too dark. He tolerated my presence—I was standing inside the bunker next to him, photographing him through the observation slit as he greeted his lady friends in passing.

For whatever reason, Willie certainly had a way with the ladies. They all liked to touch him with their trunks and greet him as they passed. This was not their usual behavior around bulls. Females generally viewed bulls as nuisances and would shoo the younger ones away at the slightest advance, often ignoring the older ones all together. But their behavior was markedly different toward their two favorites, Willie and our resident and rather solitary bull, Gakulu, our two most gentle giants.

Could it be that female elephants seemed to prefer the presence of nice guys and not the bullies? Interestingly enough, these nice guys were also good mentors for the youngsters. It didn't seem likely that that would factor into female choice, but these events did make me wonder if females were aware that their sons were going to be in good hands with these kind fellows looking after them. Up until now, it looked more like most mothers wanted to unceremoniously rid themselves of their coming-of-age sons, regardless of their social prospects once out of the family.

Meanwhile, Greg had led the rest of the boys out prior to the arrival of the females. It seemed that Greg had no interest in mingling with females and made it a point to get going before the chaos of the family groups ensued. Dominance clearly served purposes other than winning the favors of females.

I watched Willie inspect the girls and in turn allow himself to be inspected by the young bulls within the family group, and I could hear his namesake's lyrics in the back of my mind: "I've done all I can do in the name of love today." Greg was back in the driver's seat all right, but Willie was in the back seat, getting all the action.

Closure

———— ▾▴▾▴▾▴▾▴▾ ————

The 2007 season ended as it began, the boys as active as ever. Despite all the action, the elephants seemed mellower this year, probably due to the reestablished pecking order. And even though most returned to their previous ranks, in the reshuffle, Jack came out quite a bit ahead. In 2005, he ranked number twelve. This year he climbed to seventh. All of that trunk-wrapping sweet talking he did with his buddies appeared to pay off.

Kevin climbed to second rank as predicted, and interestingly enough, never showed signs of musth this season. Was it true that, in order to be a card-carrying member of the club, musth was not allowed?

Gakulu, who hadn't shown up in 2006 but had been a resident in previous years, surprised us by reappearing. In the past, he had been pretty much a loner, except for his frequent visits with his lady friends, avoiding the boys' club altogether, seemingly due to a tiff with Greg. But this year he was genuinely congenial, initiating social overtures with a trunk-to-mouth here or a body rub there. It was such an interesting role reversal, and there was no way to tell what brought it on. After all, Greg was still on top. But in 2005, he and Greg were equally able to displace one another. Was this a case of "if you can't beat 'em, join 'em?"

The only serious contention was between Willie and Prince Charles—a continuation of an old spat that we noticed in 2005. Willie won that challenge, all in a subtle tap dance around the head of the trough, Willie giving the Prince a firm foot toss across the trough every now and again to reinforce who was boss. It seemed like it would have been pretty hard to peeve such a cool cucumber as Willie, but the Prince must have done something along the way, as there was even a day the 2007 season when Willie didn't want him to come to the water hole at all.

As Willie approached the water, he kept turning around to look over his shoulder. Some distance behind him, we could see Prince Charles emerging from the bush. Willie was halfway into the clearing when the Prince broke cover and proceeded to follow Willie to the water.

Willie turned around to face him head on. Will stood as tall as he could and held his ears out. His posture said something like—"Over my dead body are you going to drink with me, son." But Prince Charles looked at Willie and kept on walking, making a semicircle around this challenger and continuing on his way. This sent Willie into a state of perturbation. He marched over and headed Prince Charles off at the pass just as he was crossing the pan to access the head of the trough. "Oh no you don't!"

But again, Prince Charles sized up the threat and proceeded to the trough. Willie stamped his feet, as if infuriated that he had had no effect on this punk. And while Prince Charles drank at the head of the trough, Willie stood staring at him—and I found myself humming, "Are there any more real cowboys in this land"? It seemed that Willie feared he was the last of a great breed.

When the sun set on the second-to-last day in camp, I had to take a moment to figure out how it was possible that the season was over so quickly. There was so much elephant activity and the camp was so busy with data collection and dung processing that there had never been a lull.

Everything seemed more intense with the prospect of departure looming large. The stars looked particularly brilliant, and I felt compelled to stay up to watch the migration of the Southern Cross across the sky one last time.

The next day was going to be big. Packing up camp was never easy, emotionally or physically. I ran through the checklist of things that I needed to do. It was a lot of work, and things had to be done in a particular order so that we remained safe throughout the process of breaking down our security perimeter and packing up our equipment.

I added a few more things to my list. Since we had built a new tower this year with all the tents on the first story, the breakdown was going to be easier. But I had forgotten to remind myself to remove the small temperature gauges from the legs of the tower. I mounted one at ground level, one at two meters, and one on the third story, eight meters off the ground. Now I needed to retrieve them.

These gauges recorded temperatures throughout the day and night during our field season so that we could see if elephant movements, communication, and social patterns were correlated with temperature changes. Based on a meteorological study conducted in the park, we knew that elephants communicated more in the evening, probably because the weather at that time made it easier for sounds to travel through the air and through the ground.

In the evening, when the temperature suddenly drops with the setting sun, the cold air forms a channel—like a two dimensional tube—allowing sounds to travel more efficiently. The same thing happens in the ocean, and blue whales use this "sound channel" (called the SOFAR channel) for their long-distance communication. This could explain why elephants tended to come in to drink after dark, as if they were more comfortable changing location when they had a clearer line of communication.

By now, the Southern Cross was falling sideways toward the horizon, like a kite forced down in a big gust of wind. Together with two stars, Alpha and Beta Centauri of the constellation Centaurus, which sat below the cross and to the left, these two star clusters told early navigators on the open sea how to distinguish true south. By drawing two lines, one line from the tip of the kite and one down through the middle of the two pointers until the two lines intersect, and then drawing that imaginary line straight down to earth from there, one could tell the direction of south very reliably. I fell asleep to thoughts of navigating the southern seas guided by the Southern Cross.

I woke up the next day to the sounds of cooking, a special last breakfast made of the local staple, pup—a porridge version of the mashed potato-like mealy meal served at dinner. Several choice toppings included butter and honey or South African cane syrup and, if there were still any left in our stash, some peanuts and raisins thrown in. Because we got so little exercise during the field season, I normally didn't eat much if any breakfast, but packing up camp required hard physical labor, so it was one of the few days that I indulged myself.

We had started the breakdown process several days earlier by packing up all of our research videotapes, all of the dung samples, and anything around camp that we wouldn't need for the last few days. That made things easier when it came to the final breakdown. One part of the team dismantled the shower, another buried the toilet, and then came the dismantling of the kitchen and dung-processing station and the packing up the recording equipment.

As we were packing up camp, Johannes, our long-time research assistant, arrived to assist in the breakdown. By the time all the trucks were packed, the sun was beginning to set again. It was going to be our last Mushara sunset of the season. We ate a quick dinner of tinned vegetable curry and watched Bent Ear and her family drink at the water hole for the last time.

Johannes and I chatted over dinner, and he asked how many more years I needed to return in order to complete my study. I looked at him and smiled. He knew the answer before I had to say anything. We were becoming closer even without language, although his English *had* been improving over the years. He smiled at me and said, "Caitlin, your success is like a tire, except that while a tire keeps on getting thinner and more threadbare, your tires keep growing thicker and thicker."

He shook his head in admiration as I tried to think of a way to fill the silence with something meaningful. All I could do was smile back.

That next morning fell into place like most mornings of our departure: people taking longer to pack up their gear than they anticipated; an unexpected number of dishes to wash and restore in the kitchen box; taping up equipment boxes to be stored at Okaukuejo; squeezing some last minute items into an inevitable box of mixed themes; the moments of hesitation and the longing to return when I hadn't even left; cursing at the volume of recycling that we never accounted for; the silence and reverent goodbyes to this very special place. Yet another great season had come to an end.

Sniffing Out Your Relatives

———— ▼▲▼▲▼▲▼▲▼ ————

Most are familiar with what goes on in a wet T-shirt contest. But what about the scientific "smelly T-shirt" contest? This clever test was designed to assess a woman's ability to discriminate body odor among men, differentiating familiar versus unfamiliar, ultimately, theoretically, making an appropriate mate choice based on smell. This test relates to my elephant studies in that I had started to wonder how males might choose their associations, whether relatedness was a factor, and whether scent played a role in how a father recognized a son when he lived outside the family group.

We all carry a scent that is indicative of a set of highly varied genes that play a critical role in the initiation of the immune response. It may seem strange, and even slightly creepy, that we can smell someone else's genes, but there is mounting evidence that genes within the major histocompatibility complex (MHC) influence body odor and thus subsequent mate choices based on the attractiveness of body odor. And we're not the only experts at this—fish, mice, and sheep also show that they, too, are skilled at sniffing out a suitable mate. But how does this relate to male elephant bonds?

Scientists have performed many permutations of the original MHC study, but the results suggest that people choose mates that have MHC genes that are dissimilar to their own—although not *too* dissimilar. The explanation of how this skill evolved in nature comes back to the idea of needing to avoid mating with Aunty Sally or cousin Joe because of the dangers of inbreeding, which can lead to a weakening of the gene pool, resulting birth defects. It also turns out that the more heterozygous (or diverse) MHC genes are, the better the ability of the immune system to deal with pathogens.

It's also important to pick a mate who is not too dissimilar from oneself, as genes that are too foreign may represent maladaptive traits as well. So, if that one T-shirt is particularly repulsive, there may be more than a gut response involved. There may be a good evolutionary reason to reject the stench.

In 1995, scientists had a group of male students of known MHC profiles sleep for two nights in identical white T-shirts. While sleeping, molecular compounds secreted from sweat glands perfumed the subjects' test T-shirts.

Back in the lab, the scientists put each slept-in smelly tee in a separate box and had women smell each of six T-shirts, three that were more similar and three that were dissimilar to their own MHC gene profile, and select the one that was most pleasing (or in this case, least displeasing). The category of most pleasing was highly correlated with sexiness and, therefore, increased the likelihood of intercourse.

In a further refinement, women who were ovulating tended to choose MHC samples that were most dissimilar, and those that were pregnant chose MHC samples that were most similar, perhaps tapping into a primal instinct for communal nesting and the desire to be around family to help raise the children. Women who were taking contraceptives also chose samples that were similar rather than dissimilar, indicating that steroids could mask the ability to discriminate smell and respond appropriately.

But what about other mechanisms that disguise one's true colors? Singles: be warned of perfumes and colognes that threaten to mask your evolutionary olfactory skills!

These studies are relevant to ours because male elephants may form bonds based on an ability to detect familial orders. And after the 2007 season, we had the opportunity to analyze the fecal DNA from a number of male associates. Since we had a very limited budget, we had to choose our subjects carefully. I selected the main members of the boys' club, including Greg, Torn Trunk, Abe, Tim, Keith, Dave, Mike, Kevin, Willie, Jack, Luke, and Prince Charles, and then a number of bulls that strayed in and out of the club, such as Gakulu, Tyler, and Congo, as well as a few bulls from Ka-meeldoring that never interacted with the club, such as Kapofi and Malan.

We then engaged in the fairly elaborate process of extracting DNA from feces and amplifying small sequences of DNA called microsatellite loci, which are repeated sequences of DNA that vary between individuals, and then sequencing each sample, which consisted of permutations in the ordering of the four nucleotide bases that make up the genome (adenine [A], thymine [T], guanine [G], and cytosine [C]). We compared their variability of sequences against a standard made from three known mother-offspring pairs that we collected and analyzed as a control for first-order relatives—that is, either mother-offspring pairs or siblings—within our population. From this comparison, we found a high level of first-order (father-son or siblings) and second-order relatives (cousins) within the boys' club as compared to the rest of the population. The bonded boys were blood relatives. They were indeed family!

We didn't expect this result because bulls were thought to move away from their natal home ranges to preserve heterogeneity in the gene pool of the population. If young males grew up and never left the area, what did that mean for the future of the genetic health of this population? Was this a result of having to fence the park and, thus, hindering the natural movements and behavior of male elephants? If so, then all of the bulls in the region would show genetic similarity. But that wasn't the case.

The question that really needed answering was how these bulls were able to identify their fathers. It made sense that brothers and cousins would recognize each other from their youth when they were still in their family units and from years of bumping into one another at the various water holes, but father and sons? Was that possible?

This is where MHC and the smelly T-shirt contest may come into play. Perhaps these bulls were able to sniff out blood relatives, who then chose to stick together. There may be an evolutionary advantage in doing so.

Some social animals behave in a way that favors the reproductive success of their relatives, even at a cost to their own survival or opportunity to pass on their genes. A classic example is the bee, where sterile females help raise their mother's next generation rather than having offspring of their own. Another example is alarm calls given by squirrels, which make them more vulnerable as prey.

These behaviors are explained by kin-selection theory, which predicts that relatedness may reduce the level of aggression within a group, leading to indirect fitness benefits for kin-favoring individuals by reducing the cost of fighting or contributing to a kin's access to resources. While this seems practical, there's a catch.

Living in groups can be a double-edged sword. A recent refinement of kin-selection theory states that the pay-offs of reduced aggression toward kin may be outweighed by competition among relatives for resources. Such is the case for the house sparrow, which shows no difference in aggression toward kin versus unrelated flock mates, nor is there any advantage of kinship in improving fighting success or rank.

Researchers did find supportive behavior between parent-offspring and sibling sparrows, so perhaps, the idea that the familial boys' club stuck together for the good of the whole wasn't completely a wash. If Greg were able to protect his progeny and provide for preferred access to the available grub, they might be that much more fit and, hence, more robust when it came time to go into musth. Ultimately, this might allow them to succeed at outcompeting others for mates, passing on more of their father's genes in their wake.

While the theorists continue the debate, at a conservative first glance, what our genetic data told us was that Greg and Tim were first-order relatives. Greg was certainly old enough to be Tim's father. And as a first-order relative, he was likely too old to be Tim's much older brother—although this was not completely out of the question if one did the calculations for this long-lived species, in which there is apparently no evidence of menopause (though my personal observations indicate that there is an age beyond which elderly females do not have calves). Nevertheless, Greg and Tim most likely wouldn't have overlapped in time within the matriarchal family.

We still had to establish cutoffs between first- and second-order relatives, but the most conservative interpretation would place Willie and Keith as first-order relatives and Keith and Tim as second order (i.e., cousins). Tim was also a second-order relative of Dave and Congo, which may explain Congo's affinity toward Tim.

Willie and Kevin ranked as cousins as well. Perhaps the incessant sparring that Kevin inflicted on Willie was part of an age-old unsettled score between them as second-order relatives that grew up in the same extended family.

Of the high-ranking bulls, Greg and Abe had the most blood relatives within the boys' club. Interesting that Abe would fall out with Greg, as seen in their later tiffs. And of the midranking bulls, Keith was related to practically everyone. Was there some connection between his bloodline and the fact that Greg was treating him more and more like a rising star?

Greg was second order to Keith, the young and ambitious Ozzie, Dave, and Luke. Also of note was that Torn Trunk was more distantly related to Greg than many of his other subordinates, and yet they were best buddies.

As for hormone levels, a pattern of higher amounts of stress during wet years, when the hierarchy was less linear, was reconfirmed owing to having had two dry years (and a stable linear hierarchy) to compare with our wet year. I had wondered whether stress levels were higher due to the lack of linearity to the dominance hierarchy and the resulting social uncertainty, as well as whether the abundance of food and water would counter such effects. Greg definitely seemed more stressed, showing more signs of vigilance behaviors, particularly considering the amount of time he spent deciding in which direction to leave, if for no other reason than to avoid Smokey. He displayed much more confidence in 2005 and 2007 when he had his posse in his pocket.

And indeed, Greg had higher cortisol levels in 2006, as did the other higher-ranking bulls. The other interesting pattern was that, while higher-ranking individuals exhibited almost equal amounts of aggression and affiliative behavior between the wet and dry years, the younger subordinate bulls exhibited more aggression in 2006—when food and water were more abundant and thus the bulls did not need to compete for resources—suggesting that, as I had suspected, a structured society is a stabilizing influence on male elephants.

I hoped that we'd get a second wet year in the near future in order to have two wet years and two dry years to compare. Climatic fluctuations might have an impact on the social structure of the boys' club, and I pondered what that might imply in relation to climate change.

In our behavioral data analysis, we were also able to show statistically that associations were much higher between members of the boys' club than between themselves and any other bulls in the Mushara area. The boys' club had some kind of mojo holding it together, and I was starting to suspect that perhaps that mojo had not only character but also genetic underpinnings.

Where Are the Boys in Gray?

▼▲▼▲▼▲▼▲▼

It was almost the last day of June, two weeks into the 2008 field season, when we weighed anchor on the great Mushara ship and set sail after our elephant subjects who were still on their wet season walkabout after late rains brought a fifty-year flood to Etosha pan. Elephants basked in the glory of their new lakefront property, wallowing in the mud and drinking of the floodwaters that filled the 2,300-square kilometer ancient lake bed with fresh water. And for much of the season, the water was even too deep for the flamingos to nest.

Although this aqueous turn of events was fortuitous for my research in the long term, it had made for another difficult and slow beginning for my students, who were eager to immerse themselves in everything elephant, including their feces. This was the first year that I was able to include a Namibian student, Kaatri Nambandi, in the research, and I was eager to give her enough work to do to get excited about the project.

But the slow beginning had been weighing heavily on all of us. "Where were those boys in gray?" became the question of the hour and then question of the day and then question of the week.

The trees had become phantom elephants and even the odd distant giraffe was mistaken for an elephant, such were our hopes that "the boys" would be returning to their dry-season home. On occasion, more than one of us had witnessed an elephant-like tree in the act of dusting. This was when I knew that we were really hurting for an elephant sighting.

It was after the rangers had reported seeing bulls at Fisher's Pan and at smaller seasonal pans near the ranger station at Namutoni that I decided it was time to do a bit of sleuthing and take the aforementioned Mushara ship on the road. I packed up the team, binoculars, cameras, identification books, and lunch and set off for a day of reconnaissance in our Land Cruiser station wagon. The open bench seating in the back allowed us all to ride together.

We took the information about where the bulls had been sighted and steered our metaphoric Mushara ship north along the northern boundary fence toward the golden Andoni Plains, lilac-breasted rollers—among the most beautiful birds in the world with their pastel plumage and long tail streamers—flashing their iridescent blue wings as they alighted from the trees lining the track as we passed. When we arrived at Andoni, the plain was so full of zebra and wildebeest, it looked like the American Great Plains not too long ago—filled with bison and plains game. Cattle egrets followed in the wake of grazing zebra, white specks in a sea of yellow.

But the Andoni Plains were fresh out of elephants. We saw plenty of elephant tracks and evidence of mud wallowing in the drying temporary pans in the Andoni area, but no live elephants.

We turned southwest to drive along the edge of the pan on the Stinkwater Peninsula. And again, along the pan's edge, there were lots of recent tracks but no boys in gray.

By late morning, we turned onto the main tourist road heading toward Tsumcor, a popular elephant stomping ground, and then headed farther south en route to Fisher's Pan. We wanted to wait to arrive at Tsumcor until the early afternoon, when elephants were most often seen at this location.

On the way to Fisher's Pan, we could see large numbers of tracks leading from the pan's edge, across the main road and heading in the direction of Kameeldoring, which would be our final stop of the day. This year's flood had apparently caused the elephant routes to shift. Since the pan was full, elephants had no reason to head the extra ten kilometers north to Mushara, and as such, the heavy traffic for the season went between Tsumcor in the south, the pan to the west, and then Kameeldoring to the east.

We stopped off at both the Klein and Groot Okevis water holes en route to Fisher's Pan to no avail, though Groot Okevi looked particularly picturesque with the unusually green vegetation surrounding the steep sloping verdant banks leading down to a small spring. With not an animal in sight, however, we moved on.

We reached Fisher's Pan just before lunch time. There'd been lots of recent elephant traffic and we could see tracks from both family groups and bulls and plenty of old dung but nothing fresh.

Although the backwater of the pan was drying up, there was still a lot of water in the pan itself, which was littered with waterbirds bobbing about on the expanse of water—geese, ducks, waders, storks, and even the elusive blue crane. But after a loop around the pan with no sign of fresh elephant activity, we decided to head to Tsumcor for lunch as we had had some good luck with family group sightings around this time in the past.

We arrived at Tsumcor with nothing but a South African shelduck in the pan. We unpacked our lunch of bread, profusely perspiring cheese, cucumber, onion, and the last coveted bit of dried mango. We waited and waited for over an hour until the crew started to get hot and fidgety. Nothing showed up, not even an antelope. It was time to see what was going on at Kameeldoring.

Kameeldoring, meaning "giraffe thorn" is similar to Mushara in that it is an artesian well that is controlled by a ball valve to regulate the flow of water. In the late dry season, it can be an unpleasant dust bowl, where wind gets channeled between the shallow sand dunes, or *umarumbas*, and blows through the clearing, making for a very harsh stay. But outside of that time, it is actually a very picturesque place, as the soil surrounding the pan contains more clay, which means that there is usually more grass in the open clearing here than at Mushara.

It also supported the *Catophractes*, or rattlebush, a silvery gray bush that masked the approach of most animals except giraffes and elephants—the tops of whose heads were visible. The drab color of the bush turned radiant and pink in the setting sun, making this a particularly beautiful place to watch the sunset.

We took a small narrow track leading into the main firebreak that ran to Mushara in the north and to Kameeldoring in the east. There were lots of fresh elephant tracks near where the sandy track paralleled the pan. All elephant roads appeared to lead to Kameeldoring. It promised to be an extraordinarily active afternoon.

We dodged the odd thorny acacia branches that the elephants had dragged onto the sandy track and made our way onto the main firebreak and then into *Catophractes* country. It was there that my Namibian student, Kaatri, spotted the first elephant of the day—a gray forehead looming above the even grayer bush line in the distance.

We got out and sat on top of the roof of our vehicle with our binoculars to get a better look. We could see a family group making its way to the water hole three hundred meters away from us. It was a beautiful sight, especially since the only other family groups we had managed to see so far this season were at night. But we had yet to see a bull. We decided to stay one step ahead of them and settle in at the water hole so that we wouldn't scare them off when they arrived.

We arrived at Kameeldoring at 3:30 P.M., just as the harsh afternoon light was starting to mellow. There were plenty of fresh elephant tracks and lots of fresh dung all around the trough and the expansive pan. So, this is where the elephants were hanging out.

We navigated around a large springbok herd and a small group of grazing zebra to place ourselves in the eastern clearing on the far side of the pan, leaving room to negotiate a quick exit if necessary. And since many rhinos, lions, and elephants were known to frequent this spot, it was important to have an exit plan. This safer position meant that our view of the elephants at the water hole would be directly into the sun, a backlit situation that was bad for photo IDs, but would make for spectacular sunset shots if the elephants were cooperative.

There was no sign of lions in the clearing, so we opened the doors to let a little more air in. It was then that I saw an elephant bull's ghostly head emerge from the scrubby horizon. It seemed to float into the late afternoon foreground like a ghost above the gray bush. I watched through my binoculars as the bull approached and hoped that it was one that I knew. Kameeldoring often was visited by bulls with which I wasn't familiar so I didn't get my hopes up.

But sure enough, as he got close, I could see his ears and knew that this elephant was indeed one that I knew. It was Tim. I saw his signature crescent-shaped cutout in the bottom of his left ear, along with a small hole just beside the cutout. The tail hair that grows in an arc on either side of his tail, creating a heart, and small broken tusks clinched the ID, and I was thrilled to finally see one of my old favorites.

I watched Tim walk in, trunk sniffing the ground as he sauntered, head swaying, feet lumbering in a smooth gait up the dusty elephant path, heading toward the water. As Tim had a good long drink, I scanned the horizon for other possible associates, as Tim was not often a loner. I looked up the slope to the east, and then into the distant flats to the west. Sure enough, a new head emerged from the horizon to the west, a much bigger head with tusks that gleamed in contrast to the drab surroundings.

The slight wind had died down and the environment was moving toward its presunset tranquil state. And there was something about a sunset over the African bush that, no matter how much I might be jaded by the humdrum of daily research life, never failed to take hold of me.

When that enormous orange ball loomed large over the horizon, the African landscape turned bright pink in the still air. Even my chalky gray elephant subjects would suddenly take on a peachy hue as they stood in the bush dusting themselves, while, for instance, patiently waiting for their mothers to deem it safe to drink at the water hole.

With Tim at the water hole and the sun edging toward the horizon, the recipe was perfect for an active elephant sunset as I could see the family group that we had sighted earlier now visible in their approach from the south, while the bull I'd noticed in the west also continued toward the water. By this time, Tim had spotted the new bull and walked confidently back to the edge of the pan to greet him. Being the mild-mannered, mid-ranking guy that he was, I assumed that Tim must have been confident about his relationship with this older bull in order for him to make such a direct, bold approach.

I couldn't make out any recognizable features on this older fellow but he was a fine bull at the height of his game with impressively long, up-curved tusks. I noted a small hole in the bottom of his right ear, but that was the only thing I could see that distinguished him, aside from the sparse hair on the outer edge of his tail and the six-inch-long hair on the inside. But there was a bend in the tail, which made him a candidate for being Marlon Brando.

Things started to take on a different mood, however, as this new bull approached Tim with ears folded in an aggressive stance. I was surprised by this posture and wondered how things were going to turn out for Tim as he was in no way a match for this bull in his prime.

Sure enough, the two bulls came to blows with heads clashing, leather squeaking, and tusks jabbing as they sparred. They held their bodies squarely aligned as they pivoted left and right in their sandy arena, the older bull gaining ground with each collision. My words flew onto the page as I took notes while trying to keep my eye on the action—phrases, thoughts, emotions describing motion, colors, smells—anything to capture the moment and help with the later rendering of this scene.

"Heads up, coming to blows, orange dust, anxious, fear, thick leather, pushing, shoving, smell of dung in air, anguish, ears folded, trunks flying, retreat, dusty lingering, sky melting, purple"—these were the words I'd scribbled down. Everything was in motion and yet remained still—the land, the mood, the animals, and the sky.

Things weren't looking good for Tim. I wondered what he had experienced this year in a social environment that appeared to be taking on the similar chaotic nature of the 2006 season—another high rainfall year. This wasn't the type of scrap that Tim would have engaged in, outside of his tumultuous encounters with Prince Charles.

The situation reminded me of the increased incidents of aggression we had witnessed in 2006 with the collapse of the boys' club. Our long-term study seemed to conform to the theory that dominance hierarchies form to minimize conflict over access to restricted resources. Because 2006 was a very wet year, there were many places to drink and thus little competition over access to water—and therefore less need for ordered social groups. In this situation, without the structure of a hierarchy in place, we witnessed more aggression between bulls.

I wondered whether the aggression we were now witnessing was indeed a reflection of the hierarchy collapsing again due to the fifty-year flood. I hoped that things would pick up at Mushara soon so that we could get a handle on the situation. Would we find that Greg had lost his hold over the boys' club for a second time? With so many places to drink, had the don lost his mojo again? And what would the situation between Greg and Smokey be?

At this point in my musings, the light started to change as the sun sunk lower on the horizon. Tim backed off and sucked his trunk, staring at his aggressor, as if waiting for a safe moment to seek some kind of reconciliation. And just as the environment had softened—the light of a descending African sun turning the harsh drab environment into a radiant pink—so, too, did the putative Marlon Brando soften as Tim approached with a submissive upturned trunk.

To my surprise, the older bull responded in kind, and the two wrapped trunks in a mutual trunk-to-mouth greeting—a successful recovery indeed from a tense beginning. And with trunks entwined, the two pushed back and forth in a spar that had taken a dramatic turn toward mutual affection, their mood mirroring the hush of the land. Leathery skin squeaked against leathery skin in a slow motion dance that had turned from combat to tandem tai chi, as the two embraced, with the older bull's trunk over Tim's head, gently pushing to and fro like the ebb and flow of a tide. Their display moved gracefully toward the left, as if choreographed to fall in line with the now flaming sun.

The family group finally decided to break cover and enter the clearing. They walked as a long line of silhouettes against an orange sky—black elephant cutouts, some with trunks up, little ones with trunks flailing and sending dust flying as they scrambled to keep up with the rest of the group. The two bulls continued to spar against the backdrop of the sunset until the sun sank below the horizon, setting the world aflame in its path as it slowly melted behind the acacia trees, pouring over a lone giraffe showing up for its first drink of the day.

A brilliant orange band wedged between the earth and pan, elephant legs now doubling in the still reflection. These moments were the kind that any photographer would salivate over—two elephant bulls sparring as silhouettes against the crimson horizon, their legs dancing a slow tango reflected off the orange pan.

And as the glow on the pan softened, the scene became dominated by the melodious chortle of the double-banded sand grouse that were trickling in, lining the pan as they bathed and stashed droplets within their downy chests to take back to their chicks. More and more grouse poured in, their volume increasing, as the elephants quietly stared at each other, now silhouetted against an inky sky adorned with the first pinpricks of stars and a sliver of the waning moon.

In the meantime, a black rhino had shuffled in and sipped at the pan opposite them. Night had settled in and it was time to return to camp. With me at the helm and Tim holding the spotlight, we headed out, ready to observe night creatures along the way. We watched a couple of bat-eared foxes trot up the road as we slowly made our way out of the *Catophractes* bushes and into the *Terminalia sericea* woodland, the shadowy trees forming a canopy overhead.

Suddenly there was a mountainous gray form in the middle of the road. It was too massive to be a patch of dust, which often lingered on the road during this hour, a sign of the change in temperature after dark, an inversion that kept particles hanging heavily above the track. I looked up through the top of the windshield and realized that it was Gary, dribbling urine wildly and fortunately way more scared of us than we were of him. Gary whisked himself off the road as fast as his stiff startled legs would carry him, leaving a fresh steaming dung sample in his wake.

After collecting our first sample of the year, we hoped that it was just a matter of time before the elephants would return. As it happened, though, at three weeks in, the season was turning out to be so slow that I had to give my student a giraffe project to work on so that she wouldn't return to Stanford empty-handed.

Notes on ear tears and tail hair patterns of elephants gave way to notes on coloration patterns of giraffe. Dominance was harder to discern between subadult male giraffes than elephants because there was so much fission-fusion going on between giraffe groups, it was hard to keep up. After doing an admirable job with the challenges at hand, we were all nevertheless relieved when Keith, the first of Greg's posse, showed up—a sign, we hoped, that we could soon return to our elephantine mission.

Since Keith was one of the least likely bulls to arrive on his own, we generally anticipated that the rest of the group would soon follow. But this time when he came on his own, I had my doubts. If one of the most social bulls in the population showed up on his own, it was a good indication that we were not about to see a busy season. Meanwhile, we returned to the giraffes and entertained ourselves with the crazy antics of the musth bull, Beckham, who had become a frequent visitor on his own.

One day, after rolling around on the ground and then sitting on the bunker and tossing his trunk over his head, Beckham came over to investigate camp, as pumped up as any bull we had seen in musth. It was late afternoon. The air was still. We contemplated turning the electric fence on, but I hated to do that during the day, when I always felt that if elephants weren't allowed to investigate their turf, they'd take their revenge at a later time. Perhaps this was a misguided sentiment, but having someone climb down the tower to turn on the fence would have made too much noise anyway.

Sure enough Beckham walked right up to the fence and took the two fingers of his trunk tip and came within centimeters of touching the fence. This caution told me that he was keenly aware of what an electric fence was all about.

He spent a few minutes inspecting the fence in this manner before circling the camp and coming around to size up the vehicles. Sometimes one loses perspective on just how large an elephant bull really is until seeing one standing next to something that provides a sense of scale. This bull loomed so large over the trucks that it was almost comical. He could have crushed them in a heartbeat, and given the way he was loitering, I couldn't help but wonder whether he was contemplating that.

On another occasion, he took his angst out on the water pump. A leak that had begun at the beginning of the season had worsened from a seep to a trickle. Beckham seemed to get impatient with the trickle and tried to push the pump over with the full force of his forehead. He clearly wasn't particularly motivated, however, or he wouldn't have had any problem completing the task.

In the middle of the night, as a few more elephants started to show up, some had moved the rocks around the pump, creating a pool of fresh water at the base. After a few more days, the rotting pipes gave way and the leak turned into a waterspout shooting out the sides, making a little separate pan next to the pump. Not only was this problematic to our research (our discrete documentation of displacements of elephants from the head of the trough was confounded by this pool of even more preferred water), but the pump was threatening to blow, which prompted the rangers to come out to help us fix it.

During the patching job, one of the old rangers told my husband Tim that he was heading up north, as a few bulls had broken through the northern boundary fence apparently in search of the grevia, a fruit similar to an apricot. He then added, "The more the elephants prune them, the more fruit there is." I was strangely saddened by this, as it caused me to think about how the elephant was no longer able to roam freely and enjoy the fruits of the land.

By mid-July, after more comings and goings of Beckham and Mike and other single musthy fellows, Torn Trunk and Dave showed up together— the first semblance of boys' club activity. It was a huge relief to see them appear on the horizon, ears flapping as they slowly lumbered down from the northwest, heading for the water hole.

After this sighting, things started to pick up with the arrival of Abe and Dave together the following day. And, a few days after that, it was Abe, Dave, and Torn Trunk. The boys' club was reconvening.

Judging from their movement patterns, it was as if something in the environment out there had cued the boys to head back to town. And, in fact, as the environment got drier, the boys began to trickle in with greater regularity, one after another. Just as our season was ending, theirs was picking up. The mismatch of our seasons was so bad that I even tried to change my plane ticket by several weeks so that I could stay longer and catch the full soap opera. Unfortunately, though, I wasn't able to make that work.

Nevertheless, I knew it was just a matter of time before Greg would show up. And indeed he did on the day that I had predicted, based on the pattern of visits from his associates over the previous few days.

Someone spotted two gray heads looming above the trees to the north. "Elephants! To the north!" Seeing that they are two very large heads, we all raced to our positions, eager to see who was going to appear, since, so far, older bulls were only showing up with younger associates. We hadn't had the same enthusiasm when seeing just one older bull, I have to say. And we were thrilled with only two adult males showing up together—a sign of how bad it had gotten.

The large forms were making fairly steady progress, the one in the lead clearly the elder, the shape of his hourglass skull giving him away, as there was only one resident bull with such a wide forehead. Finally, just as the first bull broke cover, I could see who it was.

It was Greg. The don was back in town again.

As the second bull emerged, we saw that it was Dave, a bull younger than we had first anticipated. But since Dave had been making regular appearances with Abe and Torn Trunk, I thought for sure they'd be arriving soon enough as well.

The drink they'd come for was fairly uneventful, but we were snapping pictures of Greg all the same. His tusks had grown a bit over the past year, so he didn't look quite so gnarled as he had in the previous season when both his tusks were broken almost to the base. He seemed to be in good enough spirits until about twenty minutes into their visit, at which point he stopped drinking and lay a good two-foot length of his trunk on the ground.

He froze facing west for some time, and I knew something was up. Someone was about to arrive that Greg didn't seem happy at the prospect of seeing.

I scanned the horizon. Sure enough, there they were: three more gray heads on the western horizon. Greg didn't wait to see who the party crashers were and started out in a northwesterly direction.

Dave didn't seem pleased with this situation. Having seen that Abe, then Torn Trunk, and then Keith were heading in for a cordial drinking bout, Dave seemed eager to visit with his friends. He was conflicted, however, about where his loyalties should lie. He was drinking with the don after all.

As Greg headed out, Dave trailed in his wake, stopping, turning, looking, ears out, then moving forward and stopping, turning, freezing. Greg must have sensed that Dave was lagging because he stopped, turned around, and gave Dave a good stare before continuing on.

Dave finally made his choice. He left with the don.

What was up with Greg? Abe and Torn Trunk were two of his best buddies. Since when did Greg actively avoid core members of his posse? Things must have been worse for the don behind the scenes than we thought.

Following a few more visits from Greg, and a few more disagreements with Abe—each occasion resulting in one or the other winning a challenge in equal proportions (the challenge being who would man the head of the trough, of course)—it was time to break camp again. It took longer than we had hoped, though, and on the last day we were unable to get the boma cloth down before nightfall—which meant a later departure the following morning.

After the rest of the team left, Johannes and I finished packing remaining bits. And there always seemed to be many more remaining bits than planned. Finally, Johannes took off as well, leaving me on my own. I was to spend the last night at Mushara alone.

By myself again with the bulls, just like seventeen years earlier, I was filled with a wonderful feeling of nostalgia. I stayed up late after enjoying a splendid sunset visit by, first, Captain Pickard, followed by several family groups.

I lay on the third floor of the tower, no tent, no table, no chairs, no equipment—just me, my bedroll, the night-vision scope, the tower, and the elephants, as I watched the waning moon set.

It was completely dark when I heard a strange raspy noise down below the tower. I looked down through my night-vision scope and could barely see that some elephant was rubbing on the camp fence poles below me. I refocused the lens. It was Beckham. Later, I heard the swishing of grass around the tower as Greg, Dave, and Keith encircled me as they passed—seemingly weightless, like a gentle breeze. On the final night of the season, I got to see that Greg held a smattering of the boys' club in his grasp once again.

A Case for Dishonest Signaling

—————— ▼▲▼▲▼▲▼▲▼ ——————

The musthy Beckham had extremely low levels of testosterone in 2008, well below the population average for non-musth bulls, yet he exhibited all the outward signs of musth. One possible explanation for this finding was that Beckham was engaging in "dishonest signaling."

There are few documented cases of dishonest signals in nature, which makes sense as cheating only works if it's unexpected and occurs infrequently. Take the green frog, for example. Smaller males of this species were found to purposely lower the pitch of their calls when presented with calls of larger males—lower pitch signaling a larger frog. When the calls of small and medium-sized males were played, the response calls of the frogs were not altered.

And then there's the fiddler crab that deceives both would-be challengers and mates by growing a substandard claw to replace an original one that was lost. Apparently, these competitively inferior males still attract mates in the same proportion as their more robust competitors by waving their knock-off claw just as vigorously and still bluff their brawn by growing a claw of the same length as the missing one. Though the replacement claw is weaker, it is therefore less costly to generate.

Similarly, there's a crayfish that invests all his energy in having the largest claw, not the strongest. And then there are cases of the mantis shrimp, which bluffs potential intruders into believing it can fight, even when it is at its most vulnerable, just after molting and before its new exoskeleton has hardened. I wondered whether Beckham provided the world with yet another case of dishonest signaling, displaying all the outward behavioral signs of musth without having to expend the energy to produce the testosterone that was thought to be needed to enter into, or at least to maintain, the state of musth.

Honest signaling is well documented throughout the animal kingdom, a prominent case being the jungle fowl and its red crest. Health, physical condition, and social status all affect the appearance of the crest. Thus the red crest is a direct measure of a male's good health and a strong predictor of mating success with a choosy female. While a colorful display exacts a cost—the expenditure of testosterone lowering the immune response to some extent—the level of robust health that is signaled by the brilliant crest is not significantly worsened by the transaction.

Honest signals also shape the human world. There are many signals that humans use—facial expressions, makeup, clothing, cosmetic surgery, and fast cars, for example—but since these signals are often planned, they cannot be considered reliably honest in nature. Researchers have pointed to four reliable measures of honest signaling: influence, mimicry, activity, and consistency. These terms translate into how successful one is at causing another person's pattern of speaking to match one's own (influence), copying of one person's behaviors of nodding and smiles by another within a conversation (mimicry), increasing activity levels as a measure of interest and excitement (activity), and lastly, avoiding jerky and unevenly paced actions, which would signal a lack of focus (consistency).

These examples were set against a backdrop of a speed-dating experiment, in which men and women spent five minutes chatting with a member of the opposite sex and then, at the end of each encounter, made a note of whether they'd want to exchange phone numbers with the other person. After an hour of such interaction, the tallies were presented to the participants and phone numbers exchanged among the willing.

One might expect that men would be fairly indiscriminate and women more choosy, but as it turned out, men only wanted to exchange phone numbers with women who were interested in them. How did these men know which women wanted to date them when all information was kept anonymous?

The influence of a male's scent in female choice would not explain how men knew which women would select them. This is where honest signaling may come into play, where gestures—spontaneous giggles, smiles, and animated hand movements—may have given the interested party away.

Across the animal kingdom, honest signals often have an immunological cost, as was mentioned in the example of the jungle fowl, and that cost counters attempts at cheating. For example, testosterone could be employed to stimulate sexual signals but, in doing so, weaken the immune system. Similarly, carotenoids responsible for ornamental coloration could be dedicated either to dazzling plumage or as antioxidants in support of the immune system. These tradeoffs are thought to enforce honesty in signaling.

In the context of evolution, in order for natural selection to favor honesty over cheating, the honesty of individual signalers has to be challenged every once in a while. Perhaps this could explain Greg's challenging Kevin. And maybe he caught Torn Trunk out when he cornered him in 2005, as Torn Trunk was the only other example we had recorded of a bull displaying the outward signs of musth without elevated testosterone levels. At the time, we just assumed that Torn Trunk was on the cusp of musth, but in scrutinizing the data and photos more carefully, the outward signs, particularly the urine dribbling that occurs at the height of musth, were present. Did this explain why Torn Trunk folded under pressure, while Kevin dished it back at the don with an impressive display of courage? The one honest and the other not so?

What if Smokey were to have called Beckham's bluff? Could Beckham stand the heat? Unfortunately, Beckham never appeared with anyone worthy of challenging him in order to answer that question. I needed to find a few more dishonest signalers out there to get some more definitive answers. Perhaps the exceptionally wet year having given rise to an especially low concentration of bulls in any one place allowed Beckham to get away with such low testosterone levels in musth.

Mike, in contrast, was an honest fellow and had high levels of testosterone to match his musth signaling. But for all of his high drama and testosterone (which was a lot for his usually mild-mannered self), the others just didn't seem to buy into the idea that this guy could be a serious threat.

He was around during one of those crazy nights when a young female that either was just about to go into or was actually in estrus showed up, with her family lagging behind and a bunch of excited young males in hot

Beckham has a good scratch on the bunker while curling his trunk across his face and waving his ears. These musth behaviors are thought to facilitate the broadcasting of a thick secretion from their temporal glands. Often, musth bulls appear uncomfortable, as if they have an insatiable itch that they can't scratch (perhaps due to their enlarged temporal glands and an enlarged prostate gland). This can be manifested in picking their front feet up high as if prancing, foot rubbing, tail slapping, head shaking and tossing as well as actual scratching on a surface such as the bunker. This condition most likely contributes to their overall irritable disposition and aggressive signaling.

pursuit. At the perimeter of the water hole, Mike was waving one ear and then the other, but his presence didn't seem to have any deterring effect on the younger bulls.

This behavior was in stark contrast to the testosterone-laden Smokey who, under similar circumstances, would rise out of the dust of a family group entering the clearing at dusk with only one thing on his mind—the estrus female. He'd keep up with the little cow, while all the while, swatting away her other suitors with broad sweeping trunk strokes.

Perhaps in Mike's case, the young female was not yet in estrus and to him that meant it wasn't worth fighting anyone off until the time was right. Much later that same night, I watched Mike return to the water hole around two A.M. He was drinking quietly when suddenly the young cow returned on her own. She was running in one direction, then turned and ran in another, confused and agitated, her back legs drenched. Clearly she had lost track of her family.

At first, Mike appeared startled by her arrival. But after watching her run this way and that, he approached her with a nonthreatening posture, head low and trunk tentatively outstretched, but didn't get too close. He showed no sign of being interested in her, and yet the younger bulls were driven mad by her presence. Perhaps the younger bulls were fooled by the period of false estrus that occurs before estrus. Mike, though, stood and looked at her as if he were Mr. Rogers wanting to comfort a traumatized little girl by offering her a sweater and a ride home. A highly anthropomorphic sentiment, I know, but the contrast between Mike and the other bulls was striking enough to conjure up such juxtaposition, absurd as it might seem.

As for other hormonal updates, with data from two wet seasons, it appeared that, in fact, bad boys were indeed bad without their mentors to rein them in. Subordinate youngsters had significantly higher displays of aggression in 2008 than in the dry years, just as was the case in 2006. Testosterone levels were also higher for those youngsters and subadults that exhibited early signs of musth.

Despite a decrease in overall interactions in wet years, the rate of affiliative behaviors remained stable from year to year within the bonded group. Even though younger males tend to exhibit more affiliative behaviors than older bulls, the consistent rate of affiliation measured in this study could represent an inherent need for baseline social interaction across ages and may facilitate overall stress reduction, as has been found in primates.

Meanwhile, poor Greg was working hard to keep it together in a season in which Abe constantly threatened his authority.

The Don under Fire

—— ▼▲▼▲▼▲▼▲▼ ——

The 2009 season was just about to begin. We geared up in Windhoek as always and were spending our first night in Okaukuejo with a new team and a new sense of excitement about what the season would hold. We got off to a later start this year due to the back-to-back wet years and not wanting to repeat the slow start of 2008. It was the last week of June, and we had made the right choice about timing; reports were getting back to me that Mushara was now teeming with elephants.

Each year I got more ambitious about adding additional research, as well as adding more of our own lab work, which would reduce costs elsewhere and simplify transportation logistics of the final data. This year, I planned to prepare our DNA preservation solution at the Etosha Ecological Institute at Okaukeujo prior to heading up to Mushara. Since 9/11, we could no longer hand-carry this solution on the plane over, and since luggage weight limits had been reduced from seventy to fifty pounds, it was also difficult to sacrifice that much space to a salt solution.

At the institute, after an exchange of greetings and discussion of current research and collaborations, I sent most of the team off to the water hole to enjoy the late afternoon wildlife while the rest of us busied ourselves with the unexpected and careful melting of our dimethyl sulfoxide, which had frozen during the previous extremely cold night in Windhoek. With a very high freezing temperature, it was still solid when we attempted to make up our 20 percent preserving solution.

Our task was going to take much longer than I had expected. Nevertheless, we finished just as the sun was setting, and I slipped out to the water hole to join the others and to take in the scene.

When I got there, my team reported that there was a grumpy dominant bull, persistent in holding his position at the head of the spring. Beyond the stone wall of the tourist camp, he displaced a youngster while guarding his turf. All seemed like business as usual until he suddenly got spooked by a jackal. I had never seen an elephant react to a jackal like that. I thought I must have made a mistake about what I had just observed.

I was scanning the perimeter of the water hole to see if there were any signs of lion activity, when one of my new students walked up and asked why those "dogs" (meaning jackals) out there would chase an elephant. "What? A jackal wouldn't chase an elephant," I replied with certainty. "Perhaps you saw a jackal trotting behind an elephant and the elephant got spooked for some reason? They don't like things under foot, but a jackal wouldn't normally get that close."

But my student insisted that he saw a jackal actually bite an elephant in the foot and then go on to bite other jackals. I immediately realized that he must have witnessed a jackal with rabies, which was not an uncommon occurrence in the area. I had never heard of a rabid jackal biting an elephant, but apparently elephants were not excluded from the reservoir of rabies bite victims.

As we were talking, a jackal jumped over the fence and ran into the tourist area to escape the attacking jackal. I was glad that my students had been vaccinated against rabies, as there were a great number of jackals at our field site.

When we went back to the research camp to make dinner, I reported the rabid jackal sighting to the jackal researcher, who then reported the incident to the ministry vet, who went out to find the jackal but to no avail. The occurrence of rabies waxed and waned in the park, but the incidence was apparently on the rise this season. It was hard for the staff to keep up.

Later, after dinner, despite how tired we were, we returned to the water hole at about nine o'clock in the hope of seeing the famed black rhino and elephants bathed in the floodlights of the water hole. Concerned about the rabid jackal and the long dark walk between the research camp and the water hole, I decided to squeeze everyone into one of the double-cab vehicles, with a few students sitting on the wooden-slat roof rack.

On the way, we passed an old Herero woman, in full traditional dress—

voluminous colorful patchwork hat and all—sleeping next to a smoky fire outside her government-issue concrete house. I wondered about the rabid jackal but figured she was used to such risks and preferred to slumber under the stars rather than beneath a corrugated iron roof.

When we got to the water hole, we were treated to a plethora of black rhinos and elephants. There was an appropriate hush over the crowd of tourists watching a black rhino and her calf finishing a drink. As they headed out the calcrete rocky path, their toenails clinked on the stones as they clumsily shuffled off into the night. After a short while, we were all relieved to turn in and escape the creeping cold.

The next morning we drove our caravan of vehicles to the gas station to fill up on our way out of the main camp. We met Johannes there who reported that, during the night, the rabid jackal had bitten three tourists and then bit an old man in the face while he was sleeping before it was finally dispatched. The old man, apparently, like the Herero woman I had seen sleeping next to her fire, also did not like to sleep indoors and paid the price. Having heard the noise, his son came out, grabbed the jackal, and broke its neck.

The old man and the three tourists were treated at a nearby clinic, and the jackal brain saved to be sent off to a lab to confirm rabies. The man required special treatment since a bite in the face meant close proximity to the brain.

Although I had never seen a rabid elephant and didn't think a jackal bite could penetrate elephant skin, I had heard that it was possible for an elephant to contract rabies and it occurred to me how little effect our field camp electric fence would be against such a formidable intruder. I wished that the head of research hadn't told me about the rabid lion they had taken out a few weeks back after it had bitten the front of a tourist's vehicle at a nearby water hole. As if I didn't have enough to worry about in terms of keeping my team safe.

After a little more conversation about the rabid jackal, we got on our way and drove up to the park at a leisurely pace. We knew that the camp setup was going to be much simpler, since Johannes was able to get a head start and set up the electric fence and boma cloth perimeter prior to our arrival.

He was also present when a welder came to weld a new roof on the tower, our shade cloth roof having been blown off by the high winds during the rainy season. Having the new roof and the main camp structure in place helped us get a jump start on a season that we were starting later than usual.

When we arrived at camp, a contingent of the boys was already there to greet us, led by the still reigning don, Greg. And it quickly became apparent that he hadn't settled his dispute with Abe from last season. At least they were not avoiding each other outright as they had the previous year, but Abe was assertive enough to displace the grumpy Greg from the head of the trough several times during their long drink. Their disagreement would prove to color all of the handful of visits that we witnessed in 2008.

Greg once again was on tenterhooks, another wet year wreaking havoc on his hold over the boys' club and eroding his dominion over the unruly youngsters. And in addition to groups being smaller and looser in the past two seasons as compared with drier years in the past, Greg's henchman had also changed, Torn Trunk having passed the torch to Frankie Fredericks. And for whatever reason, Torn Trunk had yet to be sighted this season. I hoped that he had been forced into retirement for being too much of a gentleman rather being the elephant rangers reported having to shoot for causing trouble on a farm outside the park. Whenever a bull wasn't sighted from year to year, such incidents were always a nagging concern.

The chaos of a younger bull in musth added to the drama of the season. It was just before sunset one evening, when Greg and Frankie made their entrance and, not halfway through their drink, the half-pint, pipsqueak little Ozzie Osborne arrived to upset the calm by being in full musth. His eyes were wild—pupils pinpricks with whites showing—and seemingly confused by a testosterone demon with a plan of its own.

Seeing a half-sized bull in musth was remarkable because bulls don't normally enter musth until they are well into their three-quarter size, in their mid- to late twenties rather than mid- to late teens. What was this little devil doing, all fired up and ready for a fight?

Ozzie's hormonal condition was not lost on Greg and Frankie. Greg backed up at the sight of this brazen little boy, while Frankie held his ears out and loomed over him on his arrival, as if to warn him of the territory he was entering and to enter at his own risk. This was Greg's dominion, and any thought of usurping it was futile.

Was it unfortunate timing or was it Ozzie's intention to join the don and his henchman at their drink? The outcome couldn't possibly weigh in his favor despite his hormonal state.

I watched as the unperturbed Ozzie walked right between the two behemoths to drink at the head of the trough, "dissing" the don as if he weren't there and didn't require the requisite "kissing of the ring" with a trunk-to-mouth greeting. And not only that, but Greg actually yielded to this possessed pubescent youth by allowing Ozzie to displace him, side-stepping to make room for the rabble-rouser.

As Ozzie stepped into the don's "throne" at the head of the trough, about to take his first sip, Greg seemed to snap out of a haze and observed this unspeakable transgression with ears held out and head held high as if to say, "What am I, chopped liver? Swiss cheese? I don't even get a hello over here? No proper trunk-to-mouth greeting for the don?"

Greg's posturing continued, but surprisingly, given how upset he was, his shoulders suddenly dropped. The don hesitantly placed his trunk to his own mouth as if too tired to deal with this vortex of testosterone. He was clearly not the don he had been when he took out the musthy third-ranking bully Kevin at the beginning of 2005.

Perhaps it was time. Perhaps he was going to step down gracefully after all. But surely not to this youth? Not here and now. That just didn't seem fitting.

Frankie wasn't going to let Ozzie off that easily and gently sparred with him in his forbidden throne, subtly pushing him down a notch. This mild reprimand seemed charitable coming from the formidable Frankie Fredericks. Perhaps he sensed that Greg was a ticking time bomb and that a gentler hand was needed.

After a few sparring moves back and forth Frankie's subtle hints didn't seem to have any impact on Ozzie, who marched right up to the don and invited him for a spar. Clearly uncertain about this brazen action, Greg seemed to humor him for a while.

But it seemed only a matter of time before I would be able to sense the locomotive engine building up steam. Then, as if the steam had finally had enough time to build to a full head, Greg blew his lid. He came at Ozzie in full combat regalia, ears folded, giant head clobbering down on Ozzie's outstretched face, the little devil embracing the engagement with gusto.

"Bring it on," Ozzie postured. "Bring it on, old man. The washed-up don should throw in the towel," his not inconsiderable jousts seemed to retort. And there he was, giving the don his best left hook, as if he hadn't a care in the world about living past this moment.

Why the huge risk? I just assumed that Greg didn't take him all that seriously, otherwise he would have wounded Ozzie but chose instead to let him off the hook. For his part, Ozzie finally appeared to exhibit reason by relinquishing the throne and leaving the don to drink in peace.

As we had seen throughout our dealings with the boys' club, young elephant bulls were not risk averse. And perhaps it was to their benefit evolutionarily to be that way. In prehistoric times, for example, risk taking was adaptive for human males, given that they had to hunt to survive and that they had to defend their families. In contrast, studies have shown women to be more risk averse than men, most probably due to the practicalities of protecting their young.

One study, which sought to test how danger was perceived in infants, found that girls, but not boys, learned to associate the sight of snakes and spiders with the frightened responses of those around them. Skeptics said the results more likely had to do with a girl's superior ability to recognize facial expressions than with an inherent difference in the fear response and risk taking later in life.

But what if the study results did, in fact, indicate an inherent difference in reaction? This pattern was certainly consistent with elephants, where females seemed more risk averse than males. For the female elephant, caution and protection of the young were of utmost importance, as was assessing the mood of others and responding appropriately.

Males, in contrast, were notoriously less cautious and, whether young or old, repeatedly go into dangerous situations, such as entering farms they knew to be risky based on previous experience. And yet they still did it for the tempting reward of corn. Conversely, these same males would indeed be concerned about entering an area that they had experienced as a hunting area.

Any hunter could tell you that the elephant knew when hunting season began. So, perhaps the rewards of the habitat within the hunting area were not as tantalizing and the risks were far greater: while one might get hurt entering a farm, one could get killed if entering a hunting area.

As I watched the young risk-taker Ozzie head out on his own, I noted that his engagement with the don and his henchman was in stark contrast to Frankie's combat a few days earlier. That battle began when an unknown full-sized bull decided to test Frankie's limits, taunting him out from the water hole to the middle of the clearing, to the eventual regret of the challenger.

Dust flew as the two goliaths hurtled at each other, massive heads raised, ears folded—first a squeak of leather, then a crack of colliding ivory. The assailant was initially undeterred and kept coming at Frankie, but Frankie didn't miss a beat. He kept doling out the blows with greater purpose than I had previously witnessed in a bull standing his ground. He pushed his opponent all the way to the edge of the clearing, the new bull losing ground with each blow. Frankie stood there for a long time, holding this new bull in a withering stare until the loser relented, headed out, and disappeared into the tree line.

After this contest, Frankie had officially inherited Torn Trunk's title of henchman who left opponents bloodied and with broken tusks in his wake. Greg had chosen well. There were only a few slight disagreements between them in an otherwise peaceful relationship. This happened twice when, after Frankie had initiated a departure, Greg gave the let's-go and headed out in a different direction, as if to signal that no one decided when and where to depart until the don had spoken.

This scenario harkened back to the disputes Greg had had in 2006 with Johannes the elephant about departure direction. And we had never seen the two together since. I imagined it took some tricky navigation to keep such a formidable henchman in line. How long could such a relationship last? Could this new henchman help extend the don's reign for yet another year?

Another interesting development within these departures was the rising in rank of Keith. On several occasions, I saw that it was Keith that initiated departure rumbles, with ears flapping and mouth open. Greg seemed to be letting Keith do a lot of the departure initiation as of late, as if he were moving him up the ladder and priming him for a top position at Greg's mafia family table. I was curious about whether Frankie had anything to say about that.

And with the progression of a new season and the changing of the second-in-command, there came other changes that one might expect in a wetter year. With smaller group sizes, and many more quarter-sized bulls unleashed on a previously seemingly well-ordered boys' club, we saw more aggression among the youngsters, as well as lots more squabbling between adult females and young bulls that were on the verge of being expelled from their families. The white tusk scars to the ribs and rumps of these youngsters spoke volumes.

Black Mamba in Camp

▼▲▼▲▼▲▼▲▼

A whirlwind is caused by the speed and power of the Black
Mamba as it moves on its path, searching for revenge.

South African legend, as told by Donald Strydom
of Khamai Reptile Park, South Africa

If you've ever heard of a black mamba it was probably in context of its
being one of the most dangerous, aggressive, and largest venomous snakes
in Africa. And for those who like to collect trivia on dangerous snakes,
though it's the second largest venomous snake in the world, next to the
king cobra, the black mamba is seven times more venomous than the cobra and can grow up to fourteen feet long.

There are plenty of deadly snakes out there in Africa that don't strike
the same fear in the heart because you'd practically have to step on them
to get bitten, such as the well-camouflaged puff adder. The puff adder
is reputed to be responsible for more fatalities than any other snake in
Africa—despite the fact that, even when it gets going, it moves like a cat-
erpillar, all chubby, sluggish, and slow.

Not so the mamba, which not only has deadly venom but is also incredibly aggressive when threatened. It can rear a third of its body up off

the ground to deliver multiple bites—a up to twelve in a row—when just one bite would be enough to kill twenty to forty grown men.

On top of this is the fact that they are the fastest moving snakes alive and have been clocked at twelve miles per hour. Having previously witnessed their speed and aggression, when one of these demons found their way into camp, I knew we had to plan our mamba extraction strategy very carefully.

Several times during the wet seasons while we were living in the Caprivi region of the country, an extremely large black mamba would cross the road ahead of us and rear up to challenge the several ton vehicle barreling toward it. The first time I saw one of these ten- to twelve-foot-long beasts rearing up above the height of a Land Rover hood I was stunned. This was apparently common behavior for the mamba, which was no ordinary snake.

A colleague told me about a friend that had been confronted by a mamba, which had reared up to the driver's side window and bit the friend on the elbow. Another colleague's friend, who was driving a tractor, was attacked by a green mamba that had reared up and bitten him in the groin. Neither survived.

If one is unfortunate enough to be bitten by one of these black-mouthed devils, the neurotoxin and cardiotoxin in its venom are paralyzing, the mamba delivering an average of about 100–120 milligrams of venom per bite. When first bitten, the victim experiences pain at the site of the bite, followed by a tingling in the mouth and extremities, then dizziness, confusion, double vision, erratic heartbeat, sweating, and a loss of muscle control.

If the victim doesn't receive prompt medical attention, symptoms progress to nausea and vomiting, shortness of breath, shock, and paralysis. It is in this state of paralysis that artificial breathing can extend the victim's life in transport until antivenom can be administered. But without antivenom, the likelihood of the victim dying is almost 100 percent, the highest rate for any venomous snake. Depending on the nature of the bite, death could occur anywhere between fifteen minutes and three hours later.

There were only two cases that I knew of in which someone survived a bite from a black mamba. One was a French foreign-aid worker in the Congo who was well versed in the art of surviving a mamba attack because he was living in a very remote area and was good at making emergency plans. After being bitten, he described in his autobiography how he had written out instructions to his staff explaining what was going to happen to him and what they needed to do to keep him alive. Fortunately, his staff was not dissuaded by the apparent death of their employer and kept up mouth-to-mouth breathing over the twenty-four-hour period that it took to get him from the jungle to the hospital in the capital city.

The second victim I had actually met. He was a South African working in Kruger National Park. When he was out on foot patrol one day, he stumbled on a black mamba and was bitten in the leg. He immediately tied his belt around his thigh and hobbled back to the main road to hitch a lift with a tourist.

He was fortunate enough to flag down a car with two little old ladies from England. When he explained his situation, he apparently did so with such calm that the severity of the situation didn't quite sink in, and they exclaimed something on the order of, "my heavens!" and kept driving below the tourist speed limit of sixty kilometers per hour. He had to explain a second time. This time the ladies got exceedingly quiet, and the one driving put the pedal to the metal. He got to the hospital just in time.

Suffice it to say that I did not take our uninvited visitor lightly. And neither did Tim. Tim was down at ground level tinkering with the electric-fence controls when he heard a swishing noise within the opened boma-cloth door to the camp. He immediately went into alarm mode. Only one thing made such a noise. I was just ascending the tower ladder when he called me over.

"Caitlin, I heard a noise over by the door," he whispered. "I think it's a snake."

We both walked over to have a look without getting too close. Sure enough, there was a very small-headed, dark olive-green snake making itself comfortable in the cuff of our boma door. I knew immediately that it was a black mamba. There was no other snake of that stature and color with such a small head. It was not yet an adult, but that was no advantage in the case of the mamba, as apparently the younger snakes haven't learned how much venom they need to use to kill their prey and pack a much stronger wallop with each bite. Not a comforting thought.

My heart started to race. In my mind, all I could see was a venomous spring-loaded viper, hurtling at its attacker. How were we going to get it out of the camp? This was an extremely dangerous situation.

We discussed the prospects.

"We could just throw a rock at it and be done with it," Tim shrugged.

"We don't have that big a rock in camp, and we could still miss."

"This is where a shotgun comes in handy."

I nodded, both of us knowing the uselessness of that statement.

We stood in silence for a moment, each lost in thought about what options we had. Tim and I weren't keen to kill anything at Mushara, we but also knew that the local rangers wouldn't think twice about killing a black mamba. "Kill or be killed" was what one had said about such matters. And yet, how could we kill the snake safely?

Using his uncanny resourcefulness in the face of adversity, Tim remembered back to an adventure book series of his youth in which two boys dispatched a poisonous snake by devising a very long noose and catching it. It just so happened that I had purchased a number of six-foot-long PVC pipes that were an inch in diameter as insulators for the electric fence, and there were a few left over. Within minutes, Tim had threaded one with our clothesline cord and fashioned a noose on one end.

We instructed everyone to remain in the tower during the retrieval operation. Much discussion ensued about the actual capture and the aftermath of the capture and all the things that could go wrong during the capture.

I was worried. "What if the noose doesn't hold? What if his neck pulls right through the pipe?"

Tim devised a break system so that this couldn't happen. And he wasn't too concerned about hurting the snake: his plan was to catch the snake safely and then kill it.

During the course of our role playing, we came to the conclusion that we needed two nooses, one as a backup. Our Namibian volunteer, Rowly Brown, was eager to serve as the backup mamba catcher. He was determined that we not kill the snake and use the nooses to transport the snake safely and set it free some distance away from camp. Since Rowly had grown up in Namibia with a keen love of nature, his vote carried some weight. Tim and I talked about it and decided that that would be best since we didn't want to get in the habit of killing things, given that we were guests of nature in this situation.

As the second noose was assembled and Tim and Rowly started to position themselves, I was worried that things would happen too fast. We needed to plan for all possible outcomes if Tim didn't manage to noose the snake on the first try.

"OK, but wait," I said. "We have to be prepared that you might miss and that he will spring toward you. Tim, you know how high they can jump. And even if he doesn't bite you, he could get into a much more difficult place to catch over by the kitchen or behind the tower."

With that, Tim and Rowly took our two dining tables, folded the legs and used them as their mamba stoppers. They slowly approached the unwitting snake with nooses and large metal table guards as it settled farther into its shady nook. Tim reached forward slowly and carefully lowered the noose over the mamba's head.

With one quick movement, the thin olive snake dangled from the long PVC pipe. Tim's hand went white with the pressure of the rope wrapped around it at one end of the PVC pipe. At the other end, the noose was cinched around a deadly neck.

From the look on his face, I knew Tim wanted to break the neck then and there and remove this threat permanently. *Nobody would try to save a black mamba.* I was sure this thought was clamoring in his head.

Nevertheless, inspired by our Nambian budding naturalist, Tim refrained from acting on his kill-or-be-killed instinct and moved so that Rowly could position the second noose. The snake wriggled as Rowly lowered the second noose over its head and pulled tightly. Success!

But now to engineer the release. How were we going to do that? There were Tim and Rowly standing side by side gripping their nooses with a black mamba dangling off the end. They were starting to have second thoughts about this live wire. How were they going to free it without getting bitten?

We decided the best plan would be for the two of them to sit on the tailgate of the truck, while I drove the truck up the road, with the mamba dangling from their six-foot nooses. As I drove slowly up the sandy track toward the northern park boundary, I tried really hard to avoid bumps and the deep craters created in the road by elephants dusting.

We made it beyond the clearing and kept going until we reached a large shady tree. I stopped the truck, and they gently slid off the tailgate and walked carefully over to the tree and released the snake.

Of course, rather than making a quick exit, they couldn't resist watching the snake's response to its release. Fortunately, it didn't attempt to strike and made a quick retreat to the shade of the tree. Tim thought it looked injured, but perhaps that was wishful thinking, since Rowly was perfectly happy with its appearance as they chatted casually about it on the way back to camp, each trying to slow their fast-beating hearts.

In retrospect, the mamba extraction was quite a major event, but things were happening so quickly that unfortunately no one took a photo of this amazing capture. At the time it was too nerve-wracking. I felt that we needed all of our wits about us and ought not be distracted with cameras. One of the team was able to video the procedure, so we did have something on record for posterity.

Tim and I spent the next few days reviewing safety protocols and familiarizing ourselves with the location of the nearest hospital in the town of Tsumeb, about a two-hour drive from the field site. We also reviewed the protocol of getting out of the park gates in the middle of the night and posted phone numbers for the rangers in charge who would have to be called in case we needed to make a midnight run to the hospital.

We all had purchased emergency medical evacuation insurance, and, in theory, a helicopter could land at Namutoni to fly a patient to Windhoek if necessary. But for a snake bite and the narrow window of time we'd have in which to respond, we knew that a drive to Tsumeb might become necessary.

We kept wondering whether we did the right thing in letting the snake live. The rangers thought we were mad, and yet the twinkle in their eyes told us there was also a new level of respect in their assessment of the elephant researchers. And they gladly took a mamba noose back to their office just in case.

But one of them reminded us that mambas were territorial. He hoped that it wouldn't return.

Our hearts sank. Return? Surely not?

Baying at a
Testosterone-Filled Moon

▾▲▾▲▾▲▾▲▾

Denebola, the brilliant tail star of the constellation Leo, sat brightly next to the orange waning moon that rose like a squashed egg yolk as I sat admiring the landscape from my hammock. Two slivers past full, the moon looked as if the top had been clipped with a nail clipper as it climbed and brightened to a luminescent yellow, casting a shadow of the tower on the silver ground now claimed by squawking crowned lapwings and accompanied by the soft hooting of the spotted eagle owl.

An hour ago, I had taken the truck out to collect the musthy Smokey's dung in the pitch-black. A lioness on the prowl made for a tense dung safari. A while later, Smokey returned to the water hole with his musthy deliberations, as a lazy lioness roared to the east. Her lover responded in an equally lazy tone, having been kept up all the previous night with her persistent copulation schedule.

Each new round had been announced by her getting up and flicking the black tuft of her tail across her lover's nose. Then the rhythmic purring had begun, then the snarls and grimaces before the explosive male orgasm, followed by the contented postcopulatory purring and rolling on her back, more likely a mechanism to channel as many sperm as possible to her fallopian tubes than the leonine version of being in the throes of bliss, enjoying an after-sex cigarette. And in order to stimulate ovulation, she kept her copulation schedule with admirable alacrity.

As a bat-eared fox cried out in the distance, I listened to Smokey drinking. Today marked his first visit to Mushara for the season. He had walked up from the south in the late afternoon and almost bumped into our dung-collecting party as we were out sampling family-group defecations, looking for parasite-load differences between age classes.

There he'd been, redefining the tree line with a curl of his trunk, prancing in objection to our presence along his path to the water hole. We quickly grabbed the boluses that we had mapped out and denoted as coming from half-sized family-group members and made a fast retreat to the truck.

Feeling a little more protected, though still vulnerable, we prepared the samples in haste. Then we drove back toward the tower and around to the south to collect dung from more family-group members, while Smokey stood at the head of the trough, curling his trunk over his head and looking at us with mouth wide open, shaking his head and cracking his ears, streaming urine all the while.

I was thrilled to see Smokey again, but disappointed that both of his gloriously splayed, perfect tusks were now broken. Last year it was the left, and now the right. It made me wonder how he had been able to keep them intact until so recently, given that he didn't shy away from a challenge. Although, I have to say that amid all of his challenges, I only saw one bull accept (a musth bull that we never saw again). Others didn't seem to want to stick around for a potential fight. In order to gain that kind of respect, some truly impressive fights must have taken place at some point.

After a long drink, he started up his displays again. First the head shake, cracking his ears across his face, then he curled his trunk over his head and across his face before letting it come to rest on a tusk, all the while with a gush of urine dribbling, followed by another long deliberation about which direction would be most appropriate for his kingly departure—that is, which direction would most likely lead to receptive ladies-in-waiting or to an unruly subject showing even a hint of not respecting his place at the top of the natural order of things. I couldn't help wondering whether a new henchman would give Greg more courage in Smokey's presence, or whether it would be a repeat performance—*capo di tutti capi* meets mere local don, who has no choice but to duck and cover.

Since Smokey was our most dramatic musth bull, it was hard to imagine how he behaved when not in musth. I wondered how long his musth period lasted. More dominant bulls were supposed to be capable of maintaining musth for several months, whereas younger bulls remained in musth for only days to weeks. I was hoping the new camera trap that I had mounted to the tower this season would help solve this mystery for at least some of the key players.

Most important, when did the don go into musth? Something told me that even in musth, he wouldn't be able to take on Smokey, but, if so, that would conflict with the prevailing theory. And since I suspected Greg himself capable of suppressing other bulls in his club from going into musth, even when he wasn't in musth himself, perhaps he had more of a hold on the population than was evident from watching his hasty retreats from Smokey.

The lions were closing in on the tower now, as the waning moon rose above the roof. There was roaring to the east, to the south, and to the southwest; perhaps they were thirsty after dining on what sounded like an oryx, judging from the death cry and the snorting of onlookers we had heard during our own repast.

After a long time at the trough without much drinking and with much posturing, Smokey stood still at the edge of the pan, body bristling, as if looking at his reflection and conjuring up the rage of his inner testosterone demons with incoherent utterances, teeth clenched, like Ben Kingsley in the film *Sexy Beast*, preparing for violence. It has long been suspected that being in the state of musth is an uncomfortable, if not painful, experience. It is no wonder that they are quick to agitate.

Smokey finally paraded out the western bypass. Again, he cracked his ears loudly as he shook his head at his hormone-inspired phantoms. Then he dusted, using his trunk to scoop up sand and plunk it down over the top of his head while waving one ear and then the other presumably to waft his musthy scent, curling his trunk over his face, mouth wide open and prancing, gushing urine from his penis sheath again. He lumbered away in his signature drawn-out, dramatic style, intermittently swinging his trunk from side to side, as if to sweep away any riff-raff who would dare to settle along his path. This was a beast to reign over all beasts.

With the pan now devoid of elephants, I was starting to notice the chill. The moon hung pale overhead as the female shelduck made a splashy landing on her return to her faithful spouse, who had recently taken to spending the day dutifully guarding the pan from the competition while she warmed eggs on the nest in the bush nearby.

Previously, the pair had spent days elsewhere, presumably incubating and guarding the nest, returning to the pan by evening and leaving again in the early morning. Each flight had been accompanied by a ritual preflight briefing between the two, a forlorn honking by the male and a raspy response by his dame, before they took to the air.

I pulled my hood up and awaited the sounds of those weighty feline tongues beating against the water. I didn't want to retire until I heard them. They were bound to slip in at any moment, considering the proximity of the roaring.

I wondered what had become our resident female, Bobtail's, yearling male cubs, which kept harassing our resident black rhino, Scratchy Mc-Splithorn, in the previous season. I assumed the two subadult males that had been hanging around this season were hers, all grown up around the collar, scruffy yet formidable. They weren't welcome when the three full-grown males were in town, the ones that greeted us on our first day in camp with the sleek female athlete in tow.

It was unclear yet whom Bobtail was avoiding with her four new cubs, but she only brought them to the pan in the very wee hours, around four A.M. The contact calls to her cubs had recently given her away, and I had been spying on her ever since.

The waning moon saw the land stilled of the harsh wind and thirsty animal traffic of the previous weeks. There was a lull to the air, only a whisper in the night as the moon got sliced thinner and thinner day by day, eventually leaving us in the darkness to ponder infinity. Yet the phase of the moon had little effect on quieting the young bull elephant rabble-rousers of Mushara in the past week. They ran toward their fate like sun-drenched immortals rattling their formidable prehensile sabers at much mightier gods than they, seemingly unaware that mortality lay just below their debaucherous hormonal urges.

A slight wind had picked up, making it too cold to remain in the hammock for much longer. The shade-cloth walls of the tower luffed in a gentle breeze. Then, finally, what I was waiting for: the carnivorous tongues hitting the water, lapping at the still black pan. The perfect send-off for my late night slumber.

Relentless Wind

▼▲▼▲▼▲▼▲▼

The next day, the wind blew for twenty-four hours straight just after the coldest night yet in camp. It battered my tent throughout the night with no letup. Bat! Bat! Bat! Bat! Battttttttt! The tent stays barely withstood the force. The tower creaked, and wind wound its way through small gaps of metal and moaned and howled. It was hard to sleep through the racket.

The following day, the wind blew the sand sideways into camp, covering everything with a gritty layer. Some areas in the back of the tower looked like snow drifts where the sand had eddied into corners. It was hard to think straight much less make a meal that didn't end up with grit in it. It was exhausting. I was exhausted. My team was exhausted. But we all endured, waiting for the torture to end.

I spent much of the day in my tent, writing and watching the wind-swept pan, looking naked without its living panorama of animals. It looked like a salt pond just before a tropical hurricane—water blowing sideways, lifting off the surface, and seeming to evaporate before hitting the ground.

That evening, when the wind finally died down, the camp breathed a collective sigh of relief, just before the hordes of thirsty elephants came pouring in, rumbling and trumpeting. Dust flew, the squeak of leather sounded with a jab to a youngster, followed by the accompanying bellow, and then the displacement shuffle between dominant and submissive family groups. It was so dark with the new moon that even through night-vision scope it was difficult to sort out the chaos of dominance between groups.

Amid the chaos, I speculated about how the nine poor shelducklings were faring out there. They had appeared that morning, guided out to the pan from the bush by their mother in between sandstorms, bravely navigating them through the jackal-infested clearing and into the safety of the pan. Their vulnerable first day of life was a crash course in how not to get eaten, caught up in a dust devil, or trampled.

To my relief, early the next morning we counted nine ducklings. They had made it through their first twenty-four hours. Calm had returned to Mushara, and after an amazing session with the boys' club in the late morning, we had our first chance to see how Smokey would handle Kevin. I thought for sure we'd see a retreat on par with Greg's, but the unwavering Kevin proved me wrong.

Kevin continued to amaze with his unflappable confidence, first, when he was in musth, toward Greg, then with the formidable Beckham the season before this one, and now on this day when he didn't back down from Smokey. Granted, there was a prickly beginning to the encounter. But it had ended with, of all things, Smokey giving him a trunk-to-mouth greeting and then Kevin allowing Smokey to inspect his penis with an affiliative ear-over-rear all the while. It was a remarkable interaction between two very imposing characters.

The next morning, Abe and Willie approached in another bout of high wind, looking very tentative. Abe spent much of his time exhibiting vigilant behaviors. We hadn't seen him overlap with Greg since the beginning of the season, and, assuming the situation was the same between him and Smokey, I was sure Abe was trying to stay clear of the loose cannon.

The two spent almost an hour at the water hole, Abe occupying the head of the trough for most of the visit. Finally, Willie approached and placed his trunk in Abe's mouth when it seemed as if he was preparing to leave. Soon after, they embarked on their respective departure paths, Willie to the north and Abe to the northwest. After they left, giving us a generous contribution of dung on their way out, I could see them to the north, standing out in the distance within the tree line, still seemingly hesitant about which direction to head and whether they were going to stick together. Finally, they moved off to the north, gradually angling toward one another until they formed a line, Abe in front.

Willie was a creature of habit as always this season. Very often he'd station himself near camp, one back leg crossed over the other, holding his head up and ears out toward camp for long periods of time. The other night as I climbed up to the third floor of the tower, I noticed a dark behemoth stationed near the southwest corner of camp. Sure enough, it was Willie once again, standing in the same position as always, like a statue, back leg cocked, ears out toward camp. He seemed genuinely curious about our activities, whether during the day or during the dark nights, with our red headlamps moving vertically and laterally on the tower.

Recently, he and Keith were in a tiff, which culminated in what looked every bit like a duel from a bad western. After a long drink with the powers that be—the dominant bulls—the two lagged behind as if there was something on both of their agendas. Willie started to head north and Keith followed, allowing a good long distance between them. When Willie reached the sandy road, he turned to face Keith. Keith stopped in his tracks in line with Willie. There they stood facing each other, but at a distance of about a hundred fifty meters. One could almost hear the jingle of spurs as they shifted their weight from side to side for a long time, as if waiting for the other to make the first move.

Perhaps, with Keith increasingly the rising star, he had begun to take liberties with Willie that Willie objected to. It seemed that the deputy saw a need to put this ambitious cowboy in his place. And after a good long stare down, Willie shook his head, cracking his ears at Keith before walking off north. With some hesitation, Keith continued to follow behind, despite Willie's occasional over-the-shoulder looks. It took a lot to rile Willie but his patience was wearing thin. Keith seemed to recognize the signs and kept his distance, as the sun hung low in the sky.

In between photography sessions with a spectacular overcast sunset and elephants parading in, dust exploding yellow overhead as they ran by, we counted herds and noted their compositions from full-sized to half to baby, documenting recognizable herd members like Bent Ear and the new comer, Crumple Ear Junior (a similar ear-cartilage deformity was found in another younger female in a different group). Bent Ear was on fire tonight, displacing herds and jabbing at an unwanted young bull with surprising force. It was getting darker and darker, so we set up our night-vision screen to watch the rest of the drama on video.

There was a commotion to the left when Scratchy the black rhino had displaced the honeymooning lions from their version of the heart-shaped bed and pushed them across the clearing to the west. The exhausted female stopped to have a drink first, and then they moved off to the west out of sight.

A little while later we heard a noise like male lions fighting over food. We tried to locate the noise, but it was just out of our range of view.

A few minutes later we heard the same noise, only the noise was moving, not in a stationary spot as it would be if related to a squabble over food. We panned around and finally saw that our resident but rare visitor, Robert, the black-maned male, had stolen the bride of his golden brother, Brad. Blondie (i.e., Brad) had been usurped, and Robert was making a threatening rumbling sound as he walked his new squeeze safely to the north, where they settled in and proceeded to engage in the noisiest copulation session that the camp had ever experienced.

The new exuberant lovers maintained a strenuous copulation schedule, with intervals of six to ten minutes, beginning with the rhythmic purring, then the signature male orgasm and girl snarl before collapsing into the silence. It was going to be a long and noisy night under a socked-in sky, with only Saturn visible through a small opening in the dense cloud cover, a cottony warm blanket enveloping the chill. A welcome change.

For whatever reason, after a few days of clouds, the boys had taken to coming in at night, and we weren't able to get our samples or record behavioral data. The days were slowing down, so I assigned the students some relevant scientific articles to present within our annual journal club to keep them on their toes and the dynamic of the camp thoughtful.

As the season wound down, I spent several hours in the hammock after dinner each night despite the cold. I was starting to feel anxious about leaving again, looking out at the Milky Way and the Southern Cross with that familiar niggling pain.

The slowing of the elephant traffic should have helped me begin my period of closure, but it hadn't. And the nights were getting colder, which was unusual, as it should have been getting warmer. The departure signals were not triggering as they should have, making the looming separation more unsettling.

My team had questions about our imminent exit, but I didn't want to talk about it, didn't want to go through the steps in my mind, because once I did, then I'd have to face the reality. I wanted to put it off for just a few more days. But I knew that wasn't fair to them. They, too, wanted to go through the process of closure and then get on with packing up.

Understandably, many were onto the next thing, while I was still clinging to this one.

On my way up the tower, I looked out at the water hole and saw Greg there with the young Tonga Boy. I was sure it was Greg, just by his posture. It was so dark in the new moon that using the night-vision scope didn't help much. I was convinced that we'd get confirmation of Greg's identity by his departure vocalization. I turned on my tape recorder just in case.

Sure enough, a few minutes later, there it was, one very long and low rumble. Since he only had one companion in tow, there was no volley, and the youngster dutifully followed along, slipping into the darkness.

With this send-off, I decided to soothe myself further with some focused time viewing the Southern Cross again. The first sliver of a waxing moon had appeared after sunset. Although it was an undeniable sign that we'd be leaving soon, it was also a welcome addition to the evening sky, a pleasing break in the darkness, a counterpoint to the splendor of the Milky Way. It was as mesmerizing as watching waves break on a sandy beach, or watching a campfire lick and crackle at hardwood, or experiencing the ebb and flow of elephant bulls coming in to drink throughout the night. And as Daniel the Stargazer, my long-time friend of seventeen years, was drinking in the pan just eighty meters away below star clusters too numerous to count, I didn't want to leave. It was about time for his visit. He tended to show up at night just at the end of every season.

When I sat up to get a closer look through the night-vision scope, there was a sudden flash as a long sparkly trail arched across the sky. The shooting star reflected so brightly on the pan that it seemed to startle the gentle mammoth-tusked giant from his foot soak.

Seeing Daniel again rekindled my worry about the elephants' future. Every year the reports of poaching and conflicts with humans were direr: something had to be done to protect the elephants and to mitigate the conflict with humans. Tolerance by those sharing land with elephants was waning, as land dedicated to elephants around the world diminished.

But who was I to judge? If I were a rancher outside of Yellowstone National Park, would I want wolves in my backyard? I'd like to think I would, but it's hard to say, not being in that position. And for many people living in Africa, one night's raid by an elephant could mean a year of starvation. That was not a tolerable situation. But the efforts of both man and elephant to maintain a foothold in Africa are inextricably linked. It was hard to say what the future held, but I remained hopeful that every little bit of research into the elephant bulls' secret world would help us to appreciate them all that much more.

The chill was becoming unbearable, so I grabbed my sleeping bag from the tent. I wasn't ready to resign my perch just yet. I was determined to milk the last nights for as long as I could.

To think that I had sat in this same position twenty years ago, inside the bunker, with just me, my hammock, my tape recorder, a microphone, and a small gas cooker to heat up tea and soup, listening to elephant bulls drinking, looking out at the Southern Cross and wondering where my life would lead. Research had become much more comfortable from a three-story tower than it had been from the bunker, but I did miss the simplicity of those days, the meditative time, just me and the elephants, their slow contemplative breathing, their slow and ponderous drinking. I waited until the Southern Cross had reached the horizon before calling it a night.

I didn't even have to look out to know who had just arrived at the water hole as I was turning in. With all the cracking of ears and squeaking of leather, I knew that the drama queen Smokey had returned for a nightcap. Then I watched him dust, the sand pouring down his mountainous form, shimmering like diamond dust in the dim moonlight.

No wonder Greg made a quick exit a while back. This was the first day that the two of them had overlapped this season. Things were changing. It would start to get warmer—the vegetation would green. And the elephants would be on the move once again.

A Deposed Don

———— ▼▲▼▲▼▲▼ ————

My dog has learned to hate the sound of zipping. As the departure date for Namibia approaches every year, and my daily patterns get more frenetic, Frodo circles ever closer at the sound of every zipper being zipped, looking for an opportunity to plunk down in my lap. Somehow seeming to weigh a thousand pounds, his eyes pleaded with me not to leave him. With a heavy sigh, he'd "ask" what could possibly take me away from his devotion.

The simple addition of a dog in my life forced me to face myself in unexpected ways, and at this instance, to ask myself whether I was doing the right thing by hanging on to this life, to my field site in Etosha National Park on a shoestring budget, and to the elephants I had come to know so well.

I took a moment away from my packing to console Frodo with some soothing scratches. I promised we'd go for a beach walk as soon as I finished with the suitcase. Just a few more things remained on the list to pack, some of which I could hear spinning in the dryer. "We'll go soon," I told him as I rubbed his belly.

With eyes full of anxiety, he laid his head in my lap and sighed again.

It was getting harder and harder to tear myself away from the creature comforts of home every summer. In fact, every year I told myself that this would be the year that we wouldn't go back. This was the year that Tim and I would spend a July sitting on a tropical island somewhere, spending the daylight hours floating among brightly colored fish that seemed not to have a care in the world.

I suddenly yearned to be neutrally buoyant, slowly breathing in and out as gentle waves washed over our suntanned backs. Denial had its curative moments—providing me just enough time to come to grips with my anxiety and embrace the excitement of seeing the drama of the latest elephant season unfold.

But my anxiety niggled into my psyche, and the night before we left for Africa, I dreamed that I was bitten repeatedly by an enormous black mamba while trying to keep it out of our field camp. I had never dreamed of snakes, nor did I necessarily fear them, but just as in all of my lion dreams of previous years, it was really the weight of the responsibility that was haunting my sleep—every venomous bite a chink in my confidence, each injection by needle-like fangs reducing my moments of conquest in the real world to defeats. And as the poison entered my veins in my dreams, my subconscious scolded me for entering a zone beyond my control yet again.

If the venom of my dreams was imaginary, Greg's apparent final undoing, after six years of being on top, was all too real, we discovered on arriving back at the research site. Five years after his battle with Kevin, in 2010, he had showed up with a gaping hole on one side of his trunk near the tip, perhaps the result of an abscess.

When he drank, he spilled half the volume of his drink at every sip. He had lost a lot of weight and spent a substantial amount of time soaking his wound after his long drinking bouts. And he was extremely grumpy, casting off friendly overtures with a crack of his ears. He barely tolerated a greeting from the older Brendan, who had approached him with the utmost care one day. Despite the earfold Greg supplied as a greeting, signaling his aggressive demeanor, Brendan persisted, and in the end Greg accepted his trunk-to-mouth hello without the head thrust or ear-cracking rejection that he had given others throughout the season.

Greg didn't accept the company of his contemporaries and elders as he convalesced, perhaps for good reason. Perhaps he was hoping to sidestep a coup and return to the throne when he recovered his strength. He only came in with his younger contingent: Keith, Tim, and Spencer, as well as some new recruits—Little Donnie, Little Richie, and Hardy Boy.

The new recruits made me wonder whether Greg might pull through this rough patch, the youngsters seemingly eager to be by his side,

Close-up of Greg's wound. Every time Greg puts a full drink into his mouth from his trunk, half of it spills out, doubling the time he needs to drink.

being fresh out of their families and looking for company. Despite his crabby mood, he seemed to know how to attract the next generation of constituents—those who he might be able to claim as comrades in a time of need.

After one of his long drinking sessions with some new young recruits, the younger bulls had long since left by the time Greg finished soaking his trunk and was ready to depart. He initiated his usual rumbling bout as he left, despite being alone. He rumbled and rumbled—his long low calls unanswered—as if engaging in an old habit that wouldn't die. I couldn't help but feel sorry for him as he stood at the edge of the clearing. Was he waiting for a buddy?

Later, I heard bull rumbles in the distance. I could tell there were two bulls vocalizing. I looked through my night-vision scope again and saw that Greg was with Keith. Perhaps the don was indeed waiting for his entourage to show up, and Keith, having already been in for a drink hours earlier, had returned to collect him.

Greg and Keith walked out together, each in turn rumbling while flapping their ears. They lumbered up a path and out of sight, just like the good old days.

As I watched them leave, I felt hopeful that he'd pull through this difficult time even if he had to relinquish his position as don. But when I arrived in Namibia for the 2011 season, all I could feel was dread that Greg might not have survived.

As we drove up from Windhoek to Etosha along the long, flat, tar road lined with acacia that stretched out to the horizon, instead of feeling the promise of another epic safari I had a pit in my stomach that wouldn't go away. Not only did I feel responsible for keeping my research team safe, I also couldn't stop worrying about Greg and what would become of his posse if he were absent. His blend of benevolence and tyranny was unparalleled in his associates. If he had died from his trunk wound, there would be so many unanswered questions.

And given how many months disappeared into preparing for this research project, Greg's dire condition added to the growing list of negatives regarding effort relative to reward. To anyone on the outside, it would have seemed so easy just to walk away, to move on with life. But I had to know what would happen next—like putting one more quarter in the slot machine, hoping for that windfall. Just one more bet, one more dune, one last wave, one more peak, or one more mountain, and for me, just one more season—as if the elephants were going to bestow on me some inner truth. And in return, I harbored the perhaps foolish hope that my behavior studies and the window they provide into their society might somehow help save them from a fate that seems all too probable to contemplate.

After staying overnight in a bush camp behind Okaukuejo water hole just inside the park, we headed east toward Namutoni rest camp. The pit in my stomach was temporarily alleviated by the spectacular sight of a large flock of flamingos lining the flooded Etosha pan near Salvadore.

Long swaths of pink and black ribboned from the shore and arched in small bursts over the shimmering ephemeral lake. We stopped at the pan's edge to have a look at a sight that was highly unusual for July. I had to restrain myself from wanting to float out on a piece of driftwood and disappear beyond the grass-studded calcrete shore—to engulf myself in the honking, blustery pink.

With Windhoek having the highest rainfall in recorded history between October of 2010 and May of 2011, I wasn't exactly sure what that would mean for our field season. Etosha had received a similar abundance of rain. In other high rainfall years, our elephant research was extremely slow as there were still plenty of other places for the elephants to drink, even at the end of June.

The only thing giving me hope for a more active year despite the rains was that, when Johannes had given me an update on the camp-building operation, he said that the elephants were coming to Mushara to drink regularly. He added that there were plenty of signs of elephant activity in the region.

This great news was countered by the fact that he had found what he estimated to be a two-meter-long Mozambique spitting cobra in the bunker. My heart skipped a beat. While all the nights I spent on my own in the bunker as a young, bright-eyed, unsuspecting researcher, it didn't cross my mind—the inevitable fate I would have succumbed to had such a snake chosen to visit then—my main concern was for the present, for the burden I felt for the safety of others.

We'd have to do something about the snake, but killing wasn't an option I wanted to pursue. Tim felt differently, having removed a black mamba from camp two years ago, with bewildered rangers incredulous that we hadn't killed it. Johannes explained that he had placed a branch into the slit of the bunker that extended to the floor inside, allowing the cobra to leave on its own accord. He tracked the trail it made in the sand to the water pump, where it now resided in the rock pile surrounding the pump, sixty meters from camp. I could live with that, though controlling the flow of the water to the pan might become a challenge if the volume of elephants was as high as the reports promised. Needless to say, Tim and I spent the next day and a half in Windhoek brainstorming cobra-proofing solutions for the bunker and for camp.

Despite the cobra inconvenience, there was no describing the relief and satisfaction I felt to arrive at an already-built field camp, after years of having to construct the perimeter before nightfall to secure us from the lions. We pulled up at noon, the only inconvenience being the baking heat and a swarm of stink bugs, the sheer volume of which was difficult to navigate. Inevitably, a good number were crushed and the stench added to the many challenges of set up.

After several hours of unpacking food, camping gear, and the remaining equipment and electronics we brought with us from the states, we were already collecting data on our first elephant bull sightings of the season. Two local residents, Gary and Keith, were at the water hole when we arrived, along with at least a hundred zebra and fifty or more giraffe and a couple of eland. Gary was not in musth as he usually was this time of year, perhaps due to the impact of shifts in dominance on musth schedules or maybe even because of a wound. Gary had been the only bull we had seen that dropped out of musth within several days of showing up with a wound to his sulcus (the skin surrounding where the tusks protrude).

Gary gave us his familiar hangdog posture as he watched us drive toward our camp. His immense head and hourglass-shaped skull loomed over us as we puttered by him, submissively requesting passage.

I was thrilled to see him but still uncertain as to how the season would progress when so much rain had fallen across the country. And Gary was not the most gregarious of bulls at the best of times, so his presence was no indication of seeing the boys' club in full swing as we had in dry years.

A while later, in strutted the musthy Smokey, swinging his trunk from side to side like a gigantic octopus limb, extending out many meters and then retracting and curling over itself and his head as if it had a mind of its own. His parade of pomp was extremely effective in chasing off an old bull I had never seen before. He even left us a dung sample, which served to jump start our hormone collection for the season, the setting up of the dung station now having a sense of urgency with a musth-bull sample waiting to be processed.

As things got more settled in the kitchen and on the ground level, we took a moment to ascend to the second floor of the tower for our first-of-the-season sundowner—a cold drink at sunset—to watch an enormous red African sun sinking beyond the horizon.

While the research team enjoyed their first wild elephant sighting with Left Hook and her family, in the back of my head was the niggling feeling that the boys' club had passed its prime. I couldn't escape the question of whether Greg was actually still alive and what his absence might mean to his pachyderm posse.

But on the second day in camp, any apprehension ended at the sight of Greg marching in from the northwest. When I saw the dusty gray tops of two large elephants emerging from the tree line, I immediately recognized Greg's distinctive skull. And as he approached, I saw the telltale ear notches and tabs dangling from his left ear. Then I identified the elephant following him: Tim, with his signature crescent cutout and hole in the middle of his left ear. Greg was not only still alive but actually looking much healthier and with one of his entourage under his wing.

A little while later, when Keith emerged from the same path, I knew that at least part of Greg's posse was still intact, even if only in fragments for now. It was clear that Keith was in a tiff with Tim, and no matter how much Tim tried to supplicate with his deference and trunk-to-mouth reassurance of his recognition of Keith's higher rank, Keith would have none of it. He'd shake his head and stamp his feet at Tim's submissive advances.

Two younger bulls greet Greg by placing their trunks in his mouth simultaneously.

Even when Greg tried to rally both of them by giving them both a gentle shove on their departure, Keith made it clear that he didn't want Tim to join them.

As they departed—Greg and Keith to the west and Tim to the southwest—Tim kept looking over to the others as if wanting to change directions and join them. But each time he did, Keith stopped and gave him an over-the-shoulder threatening look, causing Tim to continue on his own path alone, not wanting to risk an altercation with the don's current most-favored underling.

As I watched this behavior, I realized that I took it for granted that elephants behaved a certain way when they knew they were being watched. The male elephant is a master at surveying the mood of others while facing the other direction and looking at them over his shoulder. If they know they are being scrutinized, they will exhibit certain body language, such as turning around to let a following elephant know that his company is not wanted. This might be better exemplified with an example of musth bull behavior. If there are no elephants at the water hole to display to, a musth bull, more likely than not, will not display (except for Smokey who seemed to enjoy displaying with or without an audience). This disinclination to display when no other elephants are around may seem obvious, but it's always valuable when research can actually verify that it isn't happenstance, and I was happy to discover that this behavior has also been demonstrated among captive elephants.

As Greg, Keith, and Tim got to the edge of the clearing, who showed up in the southeast but another boys' club member, Dave. This pattern confirmed that even though the boys were around, things weren't going so smoothly for the club. In a dry year, when the boys' club was a tight posse, other members would normally wait for the arrival of a new buddy on the scene. Today, however, Greg and Tim came in, followed by a testy Keith. Then, after they left, Dave came through, and when he'd gone, Willie Nelson and Luke Skywalker arrived. When things were tight, all of these bulls would be at the water hole at the same time, yukking it up with body contact and gentle spars and seeming to have a grand old time together.

As I watched Greg and Keith disappear into the tree line against the horizon, I knew I couldn't be greedy. Greg was alive and the boys were all here—at least many of the key players. The season promised to be way more interesting than I had expected: not only were the fellows from the boys' club back, but there was the stream of family groups that poured in after dark, in between a visit from a rhino and her yearling calf and a couple of resident lioness Bobtail's almost full-grown sons, now sporting respectable manes.

In the darkness, a bat-eared fox cried out in mournful raspy notes, and I thought about turning in. After a full day of bull activity and of wiring the camp with solar power, there was still a lot more organizing to do to get all aspects of the research fully functioning. But tomorrow was another day—and one with a lot to look forward to.

I settled into my sleeping bag, as the crickets warmed the perimeter of camp with their melodious chirping. The sound took me back to the summers of my youth spent in New Jersey, only it was much, much colder here. Nevertheless, I felt at home again in my latest summer recess, looking out at the sinking Southern Cross, soothed by sounds of lions roaring as they passed the camp to the northeast, while reminiscing about Greg and his posse. As I dozed off, my teeth were unclenched for the first time in months.

The Don Returns

▼▲▼▲▼▲▼

I woke disoriented, transitions between nights and days starting to blur as the 2011 season progressed with mild nights bleeding into still red mornings, and my heart yearned to take on the lazy rhythm of lion roars. There was a haze in the sky from a bush fire to the east, the gray-blue wall of color giving me the sense that I was socked in and alone on the edge of an ancient inland sea.

The team had gone into Namutoni for a shower and to refill our water supply. I'd had them go earlier than normal so they'd be back in time for the bulls that I anticipated arriving before noon.

I scanned the horizon periodically, particularly to the west and northwest, expecting Greg and his posse to emerge from the bush at any moment. Sure enough, a line of gray elephantine boulders appeared out of the trees along the elephant western bypass.

As I watched six of land's largest creatures approach, it was hard to think of them as animals within a study population. Rather, I was inclined to think of them as unique individuals, well known to me, and belonging to this earth with rights to the land much like our own. And yet I was still an objective scientist, documenting, for example, that Greg (no. 22 in our bull catalog) ranked number one in the hierarchy, with Abe (no. 19) second, and Kevin (no. 40) a close third, based on the scoring of displacement events.

Here, five core members of the boys' club neared, looking very much like a family, with the younger Scotty and Hardy Boy in front today, then Dave, Kevin, Greg, and Abe. Greg positioned himself between Kevin and Abe, who had been bickering all season, with Greg shoving Kevin along, as if anticipating the inevitability of what would happen as soon as they got to the water.

As anticipated, Kevin, of course, shoved the amicable Abe out of the way, and Abe continued on to the pan for a bath as if to wait until things settled down after Greg's immediate reprimand of Kevin for having stepped out of line. I watched Greg lean in to the young Hardy Boy, and wondered what family he had come from and what kind of experience had led him to become a member of Greg's inner circle.

That night, a dim half-moon slowly rose over a white landscape while I stayed up late again after another busy day of elephants. It had gotten unusually cold, such that I hadn't braved the hammock where I usually wrote after everyone went to bed, instead confining myself to my sleeping bag in the tent.

As I tucked my sleeping bag around my little foam camp chair and settled in to write, I heard the loud sound of hacking, or retching—like a dog coughing up a hair ball. The noise was coming from just next to the tower on the east. I tried to ignore it at first since I had finally gotten warm, but then the sound came again, and then again.

I put my computer down and tiptoed in my wool socks to the edge of the tower and looked down through the scope to see a puffed up porcupine with quills erect and angry. The strange noise I heard was made by the porcupine shaking it's body, making its hollow quills rattle as a threat. I looked around for the cause of the upset, but there was no sign of life in the snow-colored clearing but the group of bulls drinking at the water hole.

The porcupine, it seemed to me, had picked a problematic spot to poise itself—right along the northeast elephant path. I looked down the path to see Keith making his way out and heading straight for the porcupine. The air was still as he padded past me, flapping his ears and emitting his let's-go vocalization. Owing to his proximity to me, the experience was more of a faint, slow throbbing in my chest rather than an auditory one.

Fortunately for the porcupine, Keith stopped short of it and waited for his friends to join him. I was no longer concerned about the porcupine, however, as I fired up my recording station and wrote down the times of vocalizations off the digital tape recorder, as well as notes about the behavior of associates. The flapping of ears, for instance, usually coincided with the vocalization, and I was building a catalog of individual let's-go calls of known individuals for eventual playback studies. The catalog would make it possible for me to determine whether club members differentiated the let's go- calls of its members from those of non–club members.

For the next hour, I recorded a long series of coordinated let's-go rumbles. These calls were so prevalent among elephants within the boys' club that I often had to remind myself that the let's-go rumble hadn't even been documented in bulls. I was eager to share this finding with others. Over the years, I had noticed that the same let's-go series of interactive rumbles that we documented in females also occurs within bonded groups of males, particularly among members of the boys' club, and most especially if Greg or his protégé Keith were present. And since one of my research goals was to demonstrate that associated bulls engaged in this interactive vocal volley on departure, I was one step closer to demonstrating that male elephants indeed exhibit ritualized bonding behaviors much as females do.

Keith took a few steps forward, causing the porcupine to curl back into a ball of formidable quills, giving Keith pause about continuing. It was just as well his companions weren't responding to his rallying call since stopping to wait for them kept him out of range of the stabbing quills. He rumbled again, in the lowest frequency vocalization described in nature (next to the blue whale), in the range of ten hertz (threshold for human hearing is around twenty to thirty hertz).

This next vocalization rallied Congo Connor. But the three others at the water hole showed no interest, leaving Keith and Congo Connor hanging their trunks over their tusks while standing next to the tower and next to the stubborn porcupine, seemingly waiting for the others to finish their drinks and join their departure. Apparently, they didn't have the clout of their don, and their buddies were going to make them wait it out.

Congo leaned over and added his trunk to Keith's, who now had two trunks hanging over his right tusk, the bottom portion of each wrapped around the other. Keith rumbled again, flapping his ears just as a matriarch did when calling her family together to leave one venue for another. This time he got a much bigger reaction out of the rest of the bulls, resulting in a series of rumbles that lasted almost a minute. Two of the three bulls were recent additions to Keith's band of associates, perhaps bonds forged while Keith awaited his convalescing don to regain enough strength to reposition himself in the boys' club. But where Keith himself would fall within the hierarchy, if it were to be reorganized, was the big question.

It was well after midnight by the time Keith, Congo, and their buddies hit the edge of the clearing to the northeast. After the elephants left, the porcupine curled back up on the elephant path, and I crept into my sleeping bag to pass the night intermittently serenaded by lion roars and lapwing squawks.

At 5 A.M., I awoke to the noise of a long low bull rumble. Then I heard the telltale trickle of Greg's sloppy drinking out of his damaged trunk. I lifted my head and saw two bulls at the water hole, one standing off to the west as if waiting for the other.

I switched on my scope and saw that Greg was indeed drinking and that Keith appeared to have been waiting for him, trying to motivate his departure with a let's-go rumble. Keith rumbled again, flapping his ears and opening his mouth as he vocalized. And after a few more rumbles, Greg stopped drinking and followed Keith out, both of them rumbling and flapping ears as they left.

I was glad to see Greg one last time before I had to leave. Despite some formidable challengers to his throne this season, I was confident that Greg continued to have the right stuff to hold his own in dominance interactions, which, considering his wound, was surprising. I witnessed several instances of Greg having to wrangle with the third-ranking Kevin. He held his own with the second-ranking Abe, even defending Abe against Kevin's blatant bullying. Just like the good old days, Greg puffed up and marched head up around the pan to shove Kevin away from the best drinking spot so that Abe could resume his usurped position. Greg was doing his best to keep the peace even with his permanent handicap. And where it counted, he also rallied the troops and, indeed, remained the don of the boys' club for at least the near future.

Scramble for Power

▼▲▼▲▼▲▼▲▼

We were racing the moon to get to Mushara before July 3, 2012, so as not to miss the elephant activity during the full-moon period. Nor did I want to miss a single opportunity to observe Greg, who, when last I saw him, was still juggling the highest-ranking position in the boys' club, despite his permanently handicapped trunk.

After four very busy days of setting up, powering the camp, and troubleshooting equipment, with elephant visits in between, our research routine was slowly emerging. There was still no sign of Greg, however. While we waited for him to arrive, there were a few unexpected activities, like having to remove a very well-fed five-foot Cape cobra from our observation bunker. Apparently our cobra-proofing last season hadn't stopped a new one from getting in and making a home in a warm, shady cement block with a constant supply of mice. Tim and I noosed it, drove it a few kilometers down the road, and released it in the shade.

The first family group showed up just as the full moon rose, followed by others into the night and throughout the day after the full moon. At more than one point, there were over a hundred elephants at the water hole.

Three days after the full moon, as the yellow waning moon rose next to the tower in the east, I was once again feeling a deep anxiety about Greg. He hadn't shown up yet, despite visits from other core members of his club.

Prince Charles had been at the water hole when we pulled up on the first day. Greg's nemesis, Smokey, arrived late morning on the following day, full of his musthy drama—swinging his trunk from side to side as he swaggered in, grabbing sand with his trunk and curling his trunk over his head, creating a sandy shower. What concerned me most was having seen Abe come in without Greg.

In 2011, Greg and Abe were inseparable. The only period they were not together by choice was in 2008, when the extremely wet year caused a splintering of the boys' club and the two of them somehow ended up in a tiff to the point of Greg stamping his feet at the sight of Abe at the edge of the clearing. The next year they had somehow made their peace and were inseparable again. In 2010, after his trunk wound, Greg was intolerant of his contemporaries, including Abe, and in 2011, they were tight once again.

After a heavy day of elephant traffic on the fifth day, still no Greg. Luke and Willie arrived from the north for a great start to an action-packed day of elephant activity, including visits from some core boys' club members. Just as Willie and Luke were on their way out, Keith put in an appearance, followed by Prince Charles and Tyler with another contingent from the west, and Gary, Yoshi, and two juniors from the south.

By the time I went to bed on the fifth day, Prince Charles had returned with a few underlings, drinking noisily. As I tried to nod off, I heard a soft let's-go volley just before the wind picked up. I wondered if I had been missing something about Prince Charles. I always thought of him as a bruiser with no interest in leading the next generation looking for a home in the world of adult male elephants. Did he have the right stuff to replace Greg, if indeed Greg were to step down, or had to step down, from the top of the hierarchy? I had always thought it would be a more mild-mannered male like the up-and-coming Keith or the wizened second-ranking Abe. I felt compelled to ponder this as the slow throbbing of Prince Charles's rumbles filled my head.

On our eighth day in camp, the wind was howling. The previous two nights had been bitterly cold, and there had been a rain shower the night before. And when Abe came in for a second time without Greg, I felt a pang of disappointment and had to accept that this might be the pattern for the season. Since Abe's not a loner, it was all the stranger for him to appear on his own.

It was possible that the fire on the nearby Andoni Plains, lit by the park managers as a controlled burn, may have caused some disruption in the boys' club. Perhaps Greg got stuck to the west of the fire and was making his way south along the pan's edge and back to the sandveld and then north to Mushara. This was my hopeful thinking—that the 2012 season wouldn't mark the end of an era. But as we were well into our second week of the field season, I was starting to fear the worst. Greg might not be with us anymore.

I didn't even want to write those words, but it had never taken him so long to show up. I was holding out for another few days in the dwindling hope that the fire detained him. But the scientist in me was already asking the inevitable: Would his disappearance spell the end of the boys' club, with no one having the right stuff to replace him? Or would Abe become the new don? Despite being second in command and a clear favorite of the youngsters, he didn't seem as interested in keeping them under his wing.

Greg had that special sauce—just the right balance of aggression and affection to hold his elephantine posse together. In spite of my fears, I wasn't ready to give up on him yet. For one thing, we'd found photos of him in our camera trap data—images taken every fifteen minutes during daylight hours throughout the year. He was at Mushara in January and March, both of these images showing him to be healthy and, even more exciting, in musth, which was confirmation that he was indeed fit, and fit enough to enter the state of musth. Since it took a lot of energy for an elephant to sustain musth, this was a awfully good sign and gave me further reason to believe that Greg was out there somewhere but just hadn't returned to Mushara for some reason.

Meanwhile, in the absence of their don, it was interesting to watch the bullies—Prince Charles and Luke Skywalker—heading two factions of boys' club members. What surprised me was that while the two of them were some of the biggest bruisers in the club, being on top of these splinter groups seemed to have somehow changed their temperaments. Luke, especially, seemed more tolerant of underlings, except for the scrapper, Spencer, whom he wouldn't let anywhere near him and who didn't dare try to approach the head of the trough with Luke at the helm. And yet he tolerated others of the junior members, allowing them to share his prime drinking spot and letting them fawn over him with a trunk over his tusk and many trunk-to-mouth salutes.

I found myself wondering if the splintering of the group might cause a few younger bulls to go into musth earlier than they would have under Greg's influence. And no sooner did I have this thought than the pipsqueak, half-pint Ozzie showed up all hopped up on testosterone again, just as he had in 2008. That, of course, was when the boys' club temporarily splintered because rain—and therefore food and water—had been abundant and there was no reason to kowtow to the don of Mushara.

Now, without the don, Ozzie was back in musth again, stirring the pot, strutting around in all of his testosterone-laden, adolescent glory, anything but shy about his intent. Poor Tim happened to be in for a drink and was the first to get the brunt of the musth storm. Little Ozzie wound his trunk up as if it were a lasso and chased Tim out of the clearing with it.

Next up was the quarter-sized bull Spock and the three-quarter-sized Spencer, who entered from the southwest. Ozzie looked right through Spock and headed for Spencer, who at first attempted to confront this unexpected adversary with head up and ears out but then quickly thought better of it, and the chase was on.

Ozzie chased Spencer twice around the camp and then around the clearing before Prince Charles arrived on the scene to take things up a notch. Ozzie positioned himself at the head of the trough and engaged in a string of dramatic musth displays with Spencer and Spock watching from a safe distance on the other side of the pan. He was dribbling urine in a full stream at this point, his ears making a cracking sound as he shook his head while curling his trunk across his brow. It was clear by Prince Charles's hesitancy to enter the clearing that he suspected something was up. I assumed that he could smell Ozzie from where he was as I certainly could smell his musthy scent from the tower.

But surely he wouldn't be concerned with this pipsqueak, even hopped up on testosterone as he was? Tim and Spencer have been known to buckle under pressure, but it didn't seem likely that a half-sized bull could be perceived as a threat to a character like Prince Charles. Yet he indeed appeared concerned. Instead of heading straight in on his normal northwest elephant path, he ditched the path and headed west, making an arc around the musthy menace and aiming for the pan.

But Ozzie caught this action while he was in mid-display and decided to march out to meet his next contestant. I was sure there would be no

In musth at an atypically early age, Ozzie demonstrates the magic of testosterone by winning a challenge against Prince Charles who is twice his size.

contest, as once Prince Charles lifted his head toward an adversary, that usually ended the conflict—all settled in the matter of a single glance. But not this time.

Ozzie came at Prince Charles with ears folded and head held almost twice his height, his pointed tusks sharp and ready to engage in battle. Prince Charles did the same, however, and when his inordinately larger size didn't seem to have any effect at all on this little devil, the Prince changed his tactic. He attempted to reach out with his trunk and wrap it around Ozzie's, as if to indicate that if he just took it down a notch, no one would get hurt.

But it was clear that this approach wasn't going to work, and Prince Charles decided on the same strategy I had seen Greg engage in: a full-on blow with his head. When Prince Charles is on fire, he's a outright steam engine and can dispatch just about any bull out there other than a few of his seniors like Abe and Mike, and as we had experienced a few days before, Beckham.

But Ozzie was on fire, and in his testosterone-amped condition with seemingly little sense of consequence, he took the blow and returned it with his own. To our surprise, the interaction ended in a chase—Prince Charles booking out of the clearing as fast as he could, with the devil dog hot on his heels. With the bush at his disposal, Ozzie incorporated a large acacia bush into his musth display, tearing at it as if rolling up his sleeves for further battle.

All of this would seem normal if Ozzie were twice his age. But he was far too young to be a competitor for mates under normal circumstances. With the hierarchy reforming free of the don at the helm, maybe this little guy was able to slip between the cracks and go unnoticed, where normally such a transgression would be suppressed in the presence of older, more dominant bulls.

It was ten o'clock at night when I heard the padding of large elephant feet passing camp and switched on the night-vision scope to see that Prince Charles had finally mustered up the courage to come in for a long-anticipated drink in peace after what had been a busy night of elephant traffic. The full moon was upon us, and the elephant numbers were increasing accordingly. As I got into my sleeping bag, I listened to the Prince drinking, as well as to the pounding sound of three hyenas loping after a springbok in hopes of a late night meal.

It was twenty minutes past midnight when I was finally ready to turn the night-vision scope off and put my head on my pillow. The wind had a chill in it, and was I cold and wanted to get comfortable. I was looking forward to stretching out my back and watching the moon for a while.

I'd set my camera on a timer to start taking time-lapse pictures beginning at dawn in the hope that Bobtail and her pride would make their way into the clearing and have a dawn romp instead of roaring out beyond the perimeter of the water hole as they had done the past few nights. I lay my head down and anticipated the clicking of my camera, which would start at 5:45 A.M.

As I fell asleep, I tried to be at peace with the passing of an era—the conclusion of Greg's position as don. But I hoped that didn't mean the passing of Greg himself. I wasn't ready to accept that such a possibility had become a reality.

The next day, I woke up with that niggling feeling again—I was starting

I monitor and record elephant vocalizations from late afternoon into the night, when the conditions are best for acoustic and seismic transmission.

to have anxiety about leaving Mushara. I didn't want to face the day, but I had to start scheduling all the things that needed to be done before we left within a week. And with only five of us left on the research team and minimal vehicle space due to a shoestring budget as always, it would be that much more critical that I timed things correctly. We needed to get the camp packed, the electric fence removed, solar panels and boma cloth down, and everything packed into a trailer to take to our storage space at the Ecological Institute, which was a three-hour drive southwest of Mushara, given our full load.

It had been a hectic season with so much extra going on in camp because of having a Smithsonian documentary film team on board. One would think that by now I'd be looking forward to a change of scenery. But I wasn't. I hadn't stepped foot away from the water hole but once to look for signs of Greg and see the extent of the Andoni fire. Other than that, I'd been here, soaking up the elephant happenings both day and night, not wanting the season to end.

I planned to send the students off to the institute to heat their dung samples so that they could be sifted and made ready for export to the states for hormone analysis. They'd bring the first load of equipment, which would start the wave of packing.

But first, the film team had a list of shots that they still needed to get of our research activities, and we had a list of unresolved interactions that we were hoping would work themselves out before we left. The first obvious question on the list: Where was Greg? At that point in time, I had to come to terms with the fact that he wouldn't turn up at all this season but wanted to know, nevertheless, what had happened to him. Second: If Greg didn't return, how would this bonded group of males fare without their don—how would the hierarchy reform, if at all, and which bull would fill the void on top?

On one of our last nights in camp, I could see in the moonlight that Keith had returned in search mode, rumbling and changing directions, rumbling and placing his trunk on the ground. He'd been in for almost two hours, during which time, Abe had joined him. And now as he left with Abe, both of them rumbling and freezing and then rumbling again, the length of their calls were a few seconds longer and more frequent than normal. Were they both keeping a vigil for Greg?

Keith lagged behind as Abe moved off. He came back for another drink before rumbling and freezing, rumbling and freezing again, making his way to the south. From their pattern of searching behavior, it seemed like they were as much in the dark about Greg's whereabouts as we were. But the fact that they, too, were searching gave me some hope that they knew more than we did—that maybe Greg was still out there somewhere.

The Royal Family

———— ▼▲▼▲▼▲▼▲▼ ————

A loud bellow rocked the still night. Three stories up in the research tower, my eyes shot open. This was not the typical sound of a young male elephant being disciplined by mom. And there were no loud scuffling noises, nor the sound of water spilling from thirsty trunks that's common when an elephant family arrives for a drink in the wee hours.

I rolled over and reached for the night-vision scope next to my pillow, training my eyes on the water hole twenty feet below my hideout. To my surprise, there were only four elephants down there—too small a number to be any of the extended elephant families we had come to observe. As I watched more closely, the oldest female swatted away the others, keeping her eyes focused on the trough. On closer inspection, I could see a missing left tusk and a *W* cut out in her left ear, barely visible without the moon, which had set hours ago.

It was Wynona from the Actor family. And I could see that her companions were her daughter, her daughter's quarter-sized calf, and Wynona's half-sized calf. But why was she not with the rest of her family?

As I watched Wynona, it was clear that something was amiss. She marched down the trough with her trunk sweeping the water, seemingly distraught. Then the explanation for the commotion popped its tiny head up and Wynona pulled a wet and very new addition to the family out of the water. Wynona had a new baby.

Big Momma and Nandi kneel down to rescue a baby that has fallen into the trough. Rescue strategies vary depending on rank, relationships, and experience.

Wynona had been looking very pregnant since the beginning of the season but I hadn't thought that we'd be lucky enough to still be at the field site when one of my favorite females gave birth, which must have happened sometime in the last twenty-four to forty-eight hours since the family's last visit. In the twenty years that I've spent summers at this water hole in Etosha National Park, Namibia, studying elephants, I had never seen a birth take place at Mushara. I assumed that the clearing surrounding the water hole was too open, making a newly born calf vulnerable to predation. Whatever the reason, the new calves we see have usually had a few days to get their footing.

It was likely that Wynona was alone because she couldn't keep up with the rest of her extended family—yet she was fortunate enough to have her immediate core family with her while she gave birth. But another scenario also fit into a hypothesis that I found myself building over the course of the season: Wynona may also have purposely put some distance between herself and her extended family in order to protect her baby. You would assume that a fragile elephant calf would be coddled by everyone in the family—caring moms, watchful aunts, and playful siblings and cousins. But in a short amount of time, it became apparent that this might not always be the case.

In all the years I've studied elephants at Mushara—first for the Namibian government, beginning in 1992, afterward as a PhD student at the University of California, Davis, and during the fifteen years I've been at Stanford University—I had never seen such aggressive elephant behavior as that exhibited toward subordinate females within elephant family groups.

Elephant family groups are matriarchal, and in the Mushara area, these groups usually comprise about fifteen individuals but can range up to thirty. Elsewhere in Africa, such as Amboseli in Kenya, groups range from two to up to twenty. Often, daughters, younger sisters, and cousins pitch in to raise others' babies—a sort of kin selection that means more family genes are passed to the next generation by pitching in to protect others that carry similar genes.

Dominance among female elephants has been described in the context of the oldest and wisest making decisions about safety, access to resources, and about where and when the group should go someplace. Ranking between families in this particular environment and context is directly related to who gets access to the freshest water to drink when multiple family groups are present.

Other researchers have described pecking orders within families as being correlated with age (or size) rather than nepotistic, or genetic, in nature. At Mushara, we started to suspect that low status might not be related as much to age as to bloodline and didn't just affect individual elephants but their offspring as well.

It made me wonder: might the opposite be true? Could high status be hereditary, creating a kind of elephant royalty? Since other researchers had established that the matriarch position was inherited by the next oldest elephant, it stood to reason that hierarchies among other individuals within the family would also be age based. But at least one study indicates that there are higher-ranking families within bond groups (made up of extended families), implying that a lineage bias, or a "queen" ethos could exist within these extended groups.

Perhaps the resource-limited environment of Etosha National Park might put more extreme pressures on elephants and cause dominance interactions to play out differently than they would where water is more ubiquitous.

The implication of female elephants living in "fission-fusion" societies is that the fission dynamic—the forces pulling them apart—is passive, that somehow the optimal number of elephants that forage and survive together (the fusion part) is determined by a natural process of families getting large enough that extended families slowly evolve looser connections and become more distantly associated (the fission part).

I was now starting to believe that the dynamic of groups fragmenting might be an active process, possibly following the direct bloodline of the matriarch, where only the highest ranking, or the "queen," and her direct descendants are welcome to hold court around the best water while the others are pushed away and perhaps forced to start their own splinter groups. Because this hypothesis went against convention, I needed to make sure I was correct about what I had witnessed.

For example, I had observed the second-ranking (and also pregnant) Susan aggressively pushing Wynona away from the water hole two days prior—the last time the family came in for a drink just before Wynona gave birth. Wynona did not appear to instigate such bullying, which culminated in a trunk slap to her retreating rear end. Based on shoulder height and back-length measures, which correlate with age, Susan was older than Wynona, which would fit current research on dominance being age-related in female elephants. But there were more subtle things at play here.

Wynona's status wasn't completely unfamiliar to me. I had been monitoring her rank and poor treatment since the 2005 season and hadn't had a way to understand the underlying motivations for what seemed like an entire family ganging up on her. This year, the rain came early, and the landscape therefore dried up much earlier than normal. With very little available water, it seemed that elephant families were competing not only with other families for access to water but also with members of their own group.

Another low-ranking female within the Actor family, Greta, was getting similar treatment, along with her new baby Groucho. The most dramatic case of this kind of treatment was toward Paula and her new calf, Bruce, from the Athlete family, who were both shunned and bullied by all the others of all sizes. Again it was clear that it wasn't just the low-ranking females who were marginalized—their calves were too.

This marginalization was in stark contrast to the treatment of new babies from high-ranking females in the Warrior family group, currently the most dominant family in our study population. I'd seen three calves of different ranks from that group happily splashing in the water hole pan together, whereas calves of low-ranking females in the other two families there were shunned by higher-ranking females and their offspring. I'd also witnessed groups of high-ranking females form coalitions to rescue a high-ranking baby who'd fallen into the trough.

In another instance, to help an inexperienced, yet higher-ranking mom in the Athlete family, the matriarch, Mia, knelt down and scooped the baby out with her trunk between its little hind legs. Crisis averted, a large contingent gathered around to soothe the wee one with Paula and her calf standing some distance away.

I wondered what would have happened if Paula's low-ranking baby had fallen into the trough instead. Would the family have rallied together in rescue, or would they have stood back and watched Paula handle the crisis on her own?

Coalitions within groups of direct blood relatives in other social animals are thought to facilitate competition with lower-ranking individuals that might not be as closely related to the matriarch. Could this be going on with elephants?

There had to be an explanation for such targeted aggressive behavior toward family members. In other places in Africa where poaching has caused a breakdown of family structure and unrelated females have merged into makeshift groups, the kind of hostile behavior that we witnessed at Mushara might make sense. But the elephants of Etosha National Park are fortunate enough not to have experienced such things. And we had photographs confirming both Paula and Wynona had been with their families for at least the past eight years, proving that they were not recent unwanted immigrants. An ill elephant might also be shunned by its family. But it wasn't likely that all three of these females—Paula and Wynona as well as Greta—and their offspring were sick.

It struck me that the higher-ranking females devoted a great deal of energy to keeping the heat on the lower-ranking ones, not to mention the coordination between individuals involved in such efforts. Perhaps the concept of optimal foraging and survival of the fittest were at work here—perhaps group sizes had to be maintained at a number that optimized the foraging opportunities of higher-ranking females and their calves in order for them to ensure the survival of the next generation.

It would make sense that an individual female's fitness would increase with an increasing group size with respect to defending against predators up until a point (what's called aggregation economy). At some point, however, a larger group size would become problematic with regard to finding enough food, particularly in dry years. The matriarchs of these three families (Actors, Athletes, and Warriors) might be ostracizing lower-ranking family members in an effort to force them out of the group to preserve the fitness of the higher-ranking and perhaps more closely related individuals—even if it takes energy in the short term to consistently antagonize subordinates and their offspring. Alternatively, this concerted effort might be in place to minimize or prevent reproduction in lower-ranking females.

By collecting DNA from feces from as many individuals and family groups as possible, I was hoping to piece together an extended family tree that would either support or further complicate my hypothesis. But the process of reconstructing that family tree would take time, at least a few more years in order to finish the data collection and perform the analyses needed. For now, all I had in front of me was the behavior, and I did my best to document it.

As the season wound down at the beginning of August, the wind started to pick up. The sky was white with the dust of Etosha Pan. Large dust devils whirled around camp—elephant dung and dry yellow grass caught up in mini-twisters as they hissed past our observation tower.

Over the course of the season, the consequences of Paula's continued rejection by the Athlete family had started to show. Not only was Paula looking exhausted but so was her little calf, Bruce. A few days earlier, they arrived at the watering hole again, and as was typical of this family, they came barreling in, desperate for a drink. This time, though, they were more interested in cooling down first and headed for the pan to take a dip. Paula, accustomed to being forced to drink in the pan while the others drank fresh water, was pushed out of the pan by one of the other adult females. Bruce's cool down was interrupted, too.

Within ten minutes, I saw Paula pushed away from the family by three of the other females, all under the watchful eye of the matriarch, Mia, who didn't miss an opportunity to trunk slap Paula if she got too close to the group. The other mothers joined forces to reject Bruce as well. And when I saw a preteen male swat at him, it struck me just how bad the situation really must have been for little Bruce.

Most disturbing was the fact that he appeared to be losing energy, refusing to get up even when his mom tried to nudge him up with her feet. And worse, he wasn't nursing. It was then that I noticed just how much smaller Paula's mammary glands were than the other lactating females—as if the stress of rejection had caused her to stop lactating. Could it be that lower-ranking cows were more stressed and thus had lower calf survival? If so, hormonal suppression could indeed be a factor within female elephant hierarchies.

Other elephant researchers had demonstrated that certain families had more calves than others during times of drought, apparently due to the better knowledge base of older individuals for finding rare resources. Research by others had also shown that more dominant families had access to better food. If more dominant families ate better, it would make sense that they might have higher overall reproductive fitness. But how did this play out between females within the same family? Did similar-aged females have the same number of calves on average, or did more distantly related family members have reduced reproductive fitness? Although other elephant researchers had suggested that dominance rank is not a predictor of female reproductive fitness, I considered whether this question needed to be revisited in this arid environment.

Reproductive suppression is well documented elsewhere in nature, through either endocrine or behavioral mechanisms or both, most notably in primates such as baboons, mandrills, and marmosets but also in wild dogs, the dwarf mongoose, and many other social species. Although reproductive suppression hasn't been described yet in elephants, perhaps, at least in tough times, the dominant female elephants in my study population and those in their direct bloodline were exhibiting intolerance toward family members that were one step removed from the queen.

Hours later, when I climbed into my sleeping bag, I worried about how Bruce would pass the night. Was Paula going to be able to protect her increasingly vulnerable little calf from predators? Despite the poor treatment, it was understandable that Paula wouldn't risk going off on her own with such a small calf. As hard as this maltreatment was for me to watch, I understood that I was most likely witnessing a natural fission of elephant families, or if not active fission, possibly the suppression of lower-ranking individuals to prevent their offspring from surviving.

Wee Hours

——— ▼▲▼▲▼▲▼ ———

Later on the same night that I'd worried about Bruce, the traffic picked up again and Keith had arrived with Abe and joined up with Prince Charles, Tyler, and Musthy Mike. Keith and Mike had a nice affiliative spar before Keith got down to drinking. The scene was much more peaceful than the day before when Beckham tried to control an unruly Prince Charles who had it in for many of the regulars—Tim, Keith, and Willie—as they approached. He folded his ears in objection, first to Tim, then to Keith, and then to Willie. And it wasn't enough to displace Keith, he pursued him all the way to the pan to make his point all the more clear—you are not wanted here.

This kind of brutish behavior had been the norm for Prince Charles over the years, and our cumulative behavioral scoring of his behaviors into a data logger allowed me to make projections. He's an aggressive bull through and through, but, he has also started paying much more attention to the younger bulls, something he hadn't done in the past. However, while Greg had always been tough on the tough guys and gentler on the youngsters, Prince Charles did not display a similar ability.

Perhaps this could explain why Beckham kept a close watch on Prince Charles and easily muscled him away from the best water whenever Prince Charles attempted to share the head of the trough with him. It's sometimes not worth giving the upwardly mobile any slack. It is probably the case that the next don would have to be deferential to his seniors if he wanted to stay in their good graces. But what was striking about Beckham was that he appeared to be keeping the bully in his place—something only Greg had done with the third-ranking Kevin when he bullied Abe. This got me wondering about Beckham's chances at becoming don, but I sensed there would be a lot more tribulation before the next don would emerge.

We had been right in the thick of these happenings the day before when thirteen bulls were at the waterhole, the cameraman on the ground shooting low to fit in all the action while I sat on the camera bracket mounted to the passenger door as the director filmed my commentary from the passenger seat of our vehicle. Things got a little hairy when Beckham gave Tim a tusking and Tim had to work hard to dodge the vehicle in retreat. But most impressive was Beckham's departure. Before leaving, he placed his trunk on the hood of the vehicle and had a good long look at us. I was prepared to make a loud noise by banging on the roof if necessary but I was reluctant to do so, as startling him could have caused him to lash out. I was watching the muscles in his trunk, waiting for any tensioning, but when he was finished with us, he simply dropped his heavy trunk and sauntered off to the north with Prince Charles trailing behind. Looking up at Beckham's massive head and flaccid trunk was an awe-inspiring moment. Sure, he could have crushed the vehicle if he wanted to, but he seemed more intent on soaking us in.

Shortly after midnight, I was ready to turn in. The wind had a chill that wouldn't quit, but I wanted to watch the moon for a while. And Smokey had returned, curling his trunk across his face, still seemingly agitated by the lingering remnants of Ozzie's musthy scent trail from several days before. Smokey followed the exact path Ozzie had used when chasing Spencer and then Prince Charles around the pan. There must be something particularly distinctive about a the scent of a young bull in musth, or at least about Ozzie, as Smokey had never really bothered with the previous whereabouts of adult musth bulls like Mike or Beckham.

Perhaps the pure novelty of Ozzie in musth caught his attention. Or perhaps Ozzie's testosterone level was so high that it was drawing Smokey's notice, much as a young snake that hasn't learned to titrate and so injects much more venom than an adult would. Now that I finally had several dung samples from Ozzie, I was eager to see how his testosterone levels compared to mature musth bulls like Smokey and Mike.

Previous studies documenting young bulls in musth in the absence of older bulls hadn't reported abnormally high testosterone levels, but something was definitely cause for concern for our resident king of musth, Smokey. And I was determined to get to the bottom of Smokey's upset, particularly after Ozzie's ceremonial unseating of Prince Charles.

On the night before our departure, I heard a jackal alarm call, alerting all Mushara residents to the presence of Bobtail and her pride to the east. This would complicate the morning packing, as well as the timing of the rolling up of the boma cloth around camp since the electric fence had been removed the previous afternoon. If we rolled up the boma cloth too soon, we would leave ourselves exposed and the remainder of the packing would be all the more difficult with lions on the prowl. Fortunately, by the time we woke up and had breakfast, the lions had moved off, allowing us to proceed with our final camp breakdown and boma removal without delay.

After two extremely hard days of packing, a broken trailer hitch, flat tires, and pushing through deep sand, we finally made it to Okaukuejo before the gates closed at sunset. When we finished unpacking all of our equipment into the storeroom and locking the door, we all let our shoulders drop in relief. We had done it.

After enjoying my first real hot shower in five weeks, I sat at Okaukuejo waterhole and stretched my sore neck. The stress of packing up had gotten the better of me, and I was slowly loosening all the kinks as I watched a hundred springbok pour into the water along with a family of gemsbok, some zebras, and some wildebeests, who were all taking a midmorning drink from this world-class waterhole.

It was hard to imagine that more than twenty years had gone by since I first sat on this very bench, and also hard to imagine how fortunate I'd been over the years to be able to spend summers at Mushara with its resident elephants. And now I had all the more to anticipate next season with the new beginnings I had so recently witnessed. I was looking forward to seeing what the future had in store for Wynona's calf, Liza, as she made her way in the world.

Although there had been many July births over the years, I felt a special attachment to this little calf, having witnessed the social vulnerability of her mother over the years and knowing almost exactly when she was born, watching her first steps and missteps, some of which ended up with her in the trough, as well as watching Wynona's caution and attention to this new addition to the family. I was also lucky enough to witness Wynona's cautious introduction of little Liza to part of the extended family, watching how many of the other youngsters came over to greet her with outstretched trunks.

Even though I can never be anything but an outsider to this elephant world, it was moments like these that made me feel privileged to have a window into their lives. The birth of Liza made me feel even more bound to this place.

I left Mushara with a sense of hope, having seen Wynona's reunion with her group and having watched Bruce (Paula's calf) receive a few overtures from the other young calves. Although still on the edge of the Athlete family, these overtures provided Bruce with an opening to the possibility of socialization, however complicated to navigate. Bruce's complex situation made me all the more curious about what a young male's existence was inside the family, and how rank and experience would influence his social opportunities later in life. Perhaps things would be okay for Bruce, in that the politics of youth might be more forgiving than those of his female elders.

The Politics of Family

————— ▼▲▼▲▼▲▼▲▼ —————

As the three-quarter moon sat high in the sky at 3:30 P.M., there was a sudden rise of dust from the edge of the southwest clearing. First, a very young bull emerged from the dense scrub forest, his body, head, and trunk bobbing as he bounced in, eager for a drink. Then, three different family groups erupted from two of the southwest paths running flat out with stiff shoulders, all trying to reach the water first. The 2013 season had completely gotten away from me. I was scrambling to make sense of current politics inside of and between families.

Within moments, positions were dictated by previously settled scores. The Actor family was relegated to a trajectory leading to the pan, while Slit Ear's family (now deemed the Goddesses) went to the head of the trough, the preferred drinking spot. This was the status until the Athletes barreled in and swiped the prime position, leaving the Goddesses in the pan and the Actors pushed off in a clump, waiting for their chance for a drink. By the look of things, no one wanted to be tusked or trunk slapped, so the less-dominant families gave way, mostly without contest.

All the formidable elephant ladies were posturing in full regalia. Ursula, the matriarch of the Goddess family, won the prize for most aggressive, holding her head up and ears out, then cracking her imposing ears as she shook her head, threatening members of other families and charging at those that didn't back down quickly enough from an invisible line that demarcated privileged access to the water source. Despite her efforts, however, the top spot was won by Mia of the Athletes.

The explicit nature of how these dominance interactions played out within and between family groups year after year drew our attention. Over the course of the season, Wynona had continued to absorb the brunt of the Actor family angst, even when the family was under duress from other higher-ranking families. The Actors were the lowest ranking of the three families that arrived for a drink. But rather than band together against the other families, a recognizable family feud immediately emerged. Wynona, her calf Liza, and the rest of her small group were forced out of the pan by high-ranking females in the Actor family, despite there being room for the entire family, leaving them standing near the clearing, waiting until it was safe to get a drink from the brackish pan after the others had left.

This observation brought me back to my question about royalty. If hierarchies within elephant families form according to age, a cousin could become the next leader, instead of the daughter or younger sister of the matriarch. If this is true, then leadership is not inherited (nepotistic) in elephants; rather, it is based more on knowledge (the oldest assumed to be the wisest). Also if true, the notion of royalty would be ruled out.

Yet, as I mentioned earlier, according to optimal foraging theory, there would be an outer limit to family size to ensure enough food for the group, and in this case drink, implying that fission might not be a passive process. If that was indeed the case, how would the lines be drawn? Two competing hypotheses are equally possible: either the direct descendants of the matriarch would get preferential treatment when it came to creating a fission, and a more distantly related individual would be marginalized until they were forced to create their own core group, or those elders with stronger social bonds (if dominance within the family is purely based on the size—that is, age—of the individual), irrespective of relatedness, band together and pressure the less preferred (subordinate) individuals out.

A recent study demonstrates that ovarian cycle length varies with social status, suggesting that status can have an influence on reproductive success—something that is well documented in other species but not yet in elephants. So when push comes to shove, is blood thicker than water with regard to status and, perhaps consequently, reproductive success?

In an attempt to address one aspect of this question, we conducted a series of focal scan observations to ask whether calves of matriarchs were treated preferentially compared to calves of subordinates. The first thing

The low-ranking Paula and her new calf Bruce are pushed out of the pan by the higher-ranking Nadia.

we could measure consistently was how far a calf strayed from its mother, taking gender and age of the calf as well as rank of mom into account.

After crunching the data, we found that calves of dominant moms stray farther than those of low-ranking moms. Could it be that calves of more dominant females are given more leeway? Do they develop more confidence because they are treated better overall? We witnessed occasions where low-ranking calves were treated aggressively for straying too far or for attempting to approach calves of higher-ranking females. If such discrimination exists between lower- and higher-ranking calves, it begs the question of how female calves of equivalent age or differing ages might treat each other as adults, given their unequal treatment within the family while growing up. Would they really forgo a prior socially defined gradient and organize themselves according to age rather than bloodline? It doesn't seem like the most likely outcome.

We were planning more detailed studies with regard to early calf development and sociality in relation to mother's rank as well as conducting relatedness studies to address these questions more thoroughly. In

The pregnant, high-ranking Susan aggressively chases off the also pregnant low-ranking Wynona, making it impossible for Wynona to drink.

the meantime, we had the opportunity to witness what happened in the aftermath of the three low-ranking births last season and the interesting solutions that three subordinate moms came up with to survive.

Wynona had the internal support group of her matriarchal line, which included her daughter and her daughter's calf, as well as Wynona's half-sized and three-quarter male offspring. These family members made it easier for Wynona to separate from the extended family when she gave birth to Liza during the previous season, and since that time, she put a clear distance between herself and the rest of the family. She still coordinated her movements with them and usually showed up for a drink during or after the time that the rest of the family was present. But she never attempted to approach the preferred drinking spot until the rest of the family left. She seemed to have worked out a solution to her poor treatment by high-ranking members of the family, and it was clear that she had started her own matriarchal group.

In the case of Greta and her yearling calf, Groucho, a half-sized female—presumably an older sister of Groucho's—kept guard over the risk-taking Groucho. Where Greta cowered in the background, Groucho's older sib rallied to his side when aggression was directed toward him by dominant family members. Given that Greta appeared to be aligned within a high-ranking faction and she never attempted to break away from the group, Groucho's good fortune and guardian angel came in the form of a six-year-old or so female.

And last, there was the case of Paula and her calf Bruce of the Athlete family. Things had looked bleak at the end of last season, and yet somehow they had made it through the year. Paula didn't have the luxury of a built-in direct offspring support system. She solved her problem by breaking away from the family completely, while adopting the support of a female from her family, Nadia, that had been aggressive toward her in the previous season, as well as an older female that I had never seen before. The assumed relative also had a yearling calf, making for a nice small coalition, within which the two calves got to socialize normally. Bruce was never seen socializing last season, and any attempt by him to approach other calves was thwarted by high-ranking females within the family. His social situation was much improved this season.

Given that elephants can live as long as humans and are considered on par in terms of cognitive capacity, it's not surprising that their social lives are incredibly complex. The structure of elephant societies is influenced by constraints on their environment, predation pressure, and the strength of affiliations, kin or otherwise, and this can change on a day-to-day basis just as in humans. Perhaps the nature of family for the elephants of Etosha is defined by different parameters than it is for those living in environments where water is ubiquitous or, alternatively, where poaching pressure is greater. Either way, it's clear that elephants can't escape the politics of family.

A New Beginning

——— ▼▲▼▲▼▲▼▲▼ ———

The white sand of Mushara glowed under an almost half-waxing moon. The high-pitched tinny call of blacksmith lapwings rose above the objections of the lanky dikkops (i.e., stone curlews). Seven eland bulls shuffled one behind the other with giant dewlaps flapping and their knees clicking like chimes as they walked, signaling their fighting ability.

It was still early, only eight P.M., but I was tired. My first team had departed that morning, and I was looking forward to a few quiet days before the next session of the 2013 season began. I wanted to force myself to stay up and write a bit, but the chatter of a busy camp lingered in my head, making concentrating difficult. The inside-the-camp chatter was joined by chatter from outside the camp—the squawking of crowned lapwings over nesting sites and the honeymooning lions just next to camp—making Mushara feel less than restful.

Despite all the chatter, I thought of the elephant, Johannes, who had shown up earlier in the day. I hadn't seen him since before Greg was wounded in 2010, recalling which brought up lots of fond memories of the good ol' days. Back in 2006, Johannes, Greg, and Torn Trunk had been inseparable—the wise men, as we called the trio. And not a visit would go by without a dispute between Greg and Johannes about who would decide in which direction they would leave the waterhole and when. This afternoon, I was amused to see how he behaved within his cohort of Abe and Keith and Beckham in the absence of Greg—making the decision to leave and then, when no one followed, returning.

It's been ten years now since I shifted my focus from acoustics and family group vocalizations to the inner workings of male elephant society and, over time, I found myself wanting to know more and more about things that are impossible to know. I even found myself wanting to leave the realm of science and reason to enter some kind of Star Trekkian scenario, in which I could speak to the elephants just as whales communicated with aliens in *Star Trek IV*. A scientist isn't supposed to have these fantastical thoughts, but I couldn't help myself. I wanted to understand the elephant mind, even though I knew I probably had no business treading there with the meager tools I had at my disposal.

When Johannes headed out to the south, after a long drink with Abe and Keith and the ornery Beckham, and realized that he wasn't being followed, he stopped and hung his trunk over his tusk for almost a half an hour. I stared at this elephantine boulder on the distant horizon and wondered what he was thinking. He seemed to be in a quandary about whether to leave on his own and risk what might be a solitary next few days or whether to come back in and try a second time to rally his companions to his itinerary. As elephants are creatures of habit, I predicted correctly that he'd come back in for one last drink and try a second time to get the two to follow him.

In reminiscing about Johannes's enduring habit of signaling a departure only to have his signal ignored, I found myself revisiting that same habit of Greg's during his weakened reign of the wet year of 2006. He didn't handle the situation with such resignation. He'd stamp his feet and shake his head at the edge of the clearing only to be ignored by even his most loyal followers, Keith and Tim. Coming back in to save face must not have been easy for either of them, but particularly Greg—the don with an uncertain throne.

Back at the waterhole, Abe and Keith had had enough of Beckham's aggressive antics at the waterhole and marched off to a dusting pan in the west, Keith flexing his penis as he passed Beckham en route to the pan. After a good thorough dusting, the two of them engaged in a strange ritual that I had seen displayed most dramatically last year at the end of the season when we were packing up camp.

Greg's closest associates spent much of the end of the 2012 season freezing in this formation with trunks on the ground often aligning along compass directions, rumbling intermittently, as if searching for their don, Greg.

Keith and Abe stood at ninety-degree angles to each other, immobile, with their trunks on the ground, as if aligning with a compass direction. Last year, Keith, Abe, Willie, and Dave had done this exact thing, aligning in four different directions. The four bulls had engaged in this behavior for over an hour, periodically shifting angles and repositioning trunks on the ground. Given that they made up Greg's inner core, again, I wondered if they had been searching for him—searching for the sign of seismic rumblings or footfalls of a passing don, elephantine songlines (à la Australian Aboriginal belief) of some kind.

They appeared so intent on this coordinated behavior, it was as if they were trying to create one giant neural network with four different sensory detector arrays and processors fully engaged in the conjuring of their don, fallen or not. Again, I reverted to my Star Trekkian fantasy and wanted the impossible to come true—that the teleporter generated by these four bulls would beam Greg back to Mushara.

As impossible as it might be for elephants to generate, coordinate, focus, and propagate an electromagnetic beam of Star Trekkian energy, a recent article on how dogs align themselves along the magnetic north-south axis to relieve themselves made me want to dig deeper into the magnetics literature for what was known about large mammals ability to detect compass directions. It turns out that cattle also align along compass directions. I wondered if elephants had this ability as well, and if so, for what purpose.

As far as synchronizing behavior to reach a specific end, what they can do with coordinated infrasonic energy in both the air and ground is impressive. By coordinating their repeated calls, with one individual's call immediately following another, we found that elephants increase how long a signal lasts by three or more times. And it is well known that the repetition of longer signals makes it easier to detect such signals at a distance. It's no wonder that elephants are masters at long-distance communication. Magnetic sensing would take things to a whole new level. I watched this strange coordinated behavior a while longer and pondered whether their sensing abilities might be even more remarkable than has been documented to date.

But it was safer to marvel on that which I already knew about elephants and it was no easy feat to localize the source of a seventeen-meter-long sound wave, given how shallow phase angles are at that length. These phase angles are too shallow for us to discriminate direction, given the space between our ears, sensors separated by only about ten centimeters. The distance between elephant ears is about half a meter, which would still make localizing low frequency sounds a challenge. Consequently, it would make sense that they might engage their front and back feet as sense organs to quadruple the distance between sensors when detecting seismic signals.

As I watched these elephants spend an inordinate amount of time focused on the ground, I couldn't help feeling a terrible loss. I realized in that moment that we hadn't appreciated the full extent of Greg's influence. Greg's character may indeed have been the scaffold that held his large posse together, and I wanted to revisit every peace-keeping tactic that the elephant don had employed to pull this off. Like the times he disciplined Kevin for stepping out of line or allowing a very young bull to suck on his tusk or sneak underneath him and be permitted to drink at the head of the trough with him. I had no idea that I'd be faced with his disappearance at this point in the study—at what seemed like the height of his political career.

It was an end of an era, and the uncertainty of the future lay ahead as to whether the don would be replaced, and if so by who. And how would the next era be defined. Despite twenty years of research, I only had the reign of one don documented so far. We appeared to be at a crossroads, and it would take some time before a new don could solidify his position and start the next chapter of the boys' club. Although Prince Charles gave it his best effort last season, it didn't seem he was able to get the foothold that was needed to keep the top position and maintain a constituency. Luke tried his best this season to rally members of the junior boys' club—those that had recently left their families and joined the band of merry men—but didn't appear to have the respect of his elders needed to pull it off.

Beckham, by way of contrast, could have the whole place eating out of his trunk if he so chose. But he hadn't yet figured out how to contain his bullying ways with such well-respected members of the club as Abe and Keith. In fact, there was more penis flexing in vexation during his recent interactions with them than I had ever seen within one session, outside of Greg trying to put Kevin in his place for harassing Abe. Watching an inordinately large and engorged, veiny prehensile phallus flexing aggressively at an offender was like watching a man in a muscle shirt with a tattooed arm shaking a fist at someone who cuts him off in traffic. The indignity of it all—to challenge one's manhood over a stoplight or, in this case, a drink at the waterhole.

Certainly there are examples of characters that have defined periods of human history with their either maniacal or benevolent ways. Did Greg's character play a role in defining the behavior and dimensions of this social group? The lack of a very large cohesive group in the two dry years since his absence forced me to ask two things: First, had the boys' club without their don become simply just a smattering of semisocial individuals that had a loyal buddy or two but in general merely tolerated larger social experiences at the waterhole? And second, could that explain why such a large group of tightly bonded males was not very common in nature?

Loosely associated groups of male elephants had been reported in other areas of Africa, including tighter bonds between two or a few individuals, but was it possible that the tightknit nature of the boys' club could be explained by the influence of the character of the dominant individual?

We had witnessed why it was in a bull's best interests to kowtow to the current don in order to get a drink in peace. But would dry years in the future ever yield the same large groups and tight associations without such a charismatic male at the helm?

Could it be that there would be no replacing Greg? Was he truly a rarity in nature? Would Beckham learn to be less volatile? Could Prince Charles maintain the princely foothold that he had demonstrated in the 2012 season? Would Luke Skywalker figure out how to conjure up a Jedi hold over his elders? Greg's character suggested that high rank required both knowing just when and how to employ both the carrot and the stick, along with revering certain of his elders.

The don's seat was up for grabs, but in the meantime, at least there was the satisfaction of closure on a few things. Ozzie, for instance, turned out not to have testosterone levels that were anything out of the ordinary for a musth bull. Perhaps a unique pheromone was involved in telling Smokey that this young brazen whippersnapper was up to no good. Collecting the volatiles (organic compounds that can be detected by scent) from the dung of musth bulls might help us understand Smokey's consternation over Ozzie's flamboyant musth behavior. Smokey never became so agitated when Mike or Beckham was in musth. But perhaps character is again at play. Mike in musth is not the musthy steam train that Ozzie is. Even testosterone doesn't seem to be a cure for Mike's tentative ways. Although Beckham is not a mild-mannered bull so that rules that idea out.

Another thing closer to resolution from last season is Wynona's quasi independence from the Actors. She figured out how to operate on the periphery, just enough to provide social opportunities for her new calf, Liza. Liza had the chutzpah of a native New Yorker, greeting the old Captain Kirk towering above, navigating her peers and elders within the extended family, knowing to avoid a jagged poke from overbearing Susan, and dealing with the exuberance of the unstoppable Groucho as he sat on her during a mud bath.

Even Paula and Bruce came through their difficult year, albeit with a slightly different solution. We were thrilled to see them each time they came in, Bruce having a good frolic and dust bath with his new buddy, looking healthier than ever. All appeared well in elephant country again.

Knowing what I now know about elephants, it's hard to remember back to the distance I felt between myself and them in the early days living in the Caprivi region of the country. But the elephants of the Caprivi are a much different animal from the giants of Etosha—the difference between a society on the brink of uncertain human politics and one that still remembers how things had always been and were expected to continue, if only within the reduced range of Etosha National Park. Just as an environment defines a person and his culture, so, too, apparently for the elephant.

Meanwhile, the soap opera of the boys' club would carry on whether or not I was there to witness its members in a rapidly changing social landscape, to document the shifting of politics from year to year. I just

Greg has a bonding moment with the new recruit, Hardy Boy, demonstrating his soft touch with the youngsters.

hoped that I'd still be able to be a part of it somehow, to continue my vigil for as long as I could find a way to support it and for as long as I was welcome back.

The chill of late July had finally settled in. I lay awake listening to the retreating sounds of the seven sated eland bulls—the resonant chiming of knees in the crisp air slowly and definitively denoting the passing of the night as they disappeared into the distance.

A little while later, a slight breeze batted at the camp walls, once again sounding like the luffing of sails. I was transported back to our great Mushara ship again, the clanging of the halyard on the great mast followed by a more steady wind keeping me awake in a ship bucking to embark on an uncharted course.

I turned the ship in my mind's eye, letting the sails out in a full run heading south after the testosterone-crazed Ozzie and the haggard-looking Vanessa and her exhausted calf, Renee. Perhaps I could sweep Vanessa away from her teenage nightmare suitor, Ozzie. Or at least rescue her calf from her unfortunate predicament.

It was hard to imagine that the imposing Vanessa, who seemed to play the role of secretary of state within the Actor's family, would view Ozzie as a suitable mate. And that the musthy Smokey, at least twice Ozzie's age and stature, wouldn't have been preferable. If female choice were really at play, it was certainly not very evident. Too many unanswered questions tugged at my mind, but I was consoled by the fact that estrus only lasted for a few days—perhaps for very good reason, as it was exhausting to watch the disruption to Vanessa and her family.

I imagined steering our great Mushara ship on into the silver night. On fickle winds, the Southern Cross continued to guide my search for resolution as to who would become the next elephant don.

ACKNOWLEDGMENTS

———— ▼▲▼▲▼▲▼▲▼ ————

The research described in this book would not have been possible without the continued support and collaboration of my dearly beloved partner in most crimes, Tim Rodwell, the enthusiasm, dedication and bull ID addiction of my MS student at the time, Colleen Kinzley, and the intellectual interest and analytical eye of my former postdoctdoctoral fellow, Jason Wood, and discussions with mentors Robert Sapolsky, Sam Wasser, and Sunil Puria. The research also benefited from conversations with collaborators and colleagues including Francis Steen, Frans deWaal, Robert Jackler, Wendy Turner, Donna Bouley, Maggie Wisniewska, Barbara Durrant, Kathleen Gobush, Rebecca Booth, Werner Kilian, Wilferd Fersfeld, George Wittemyer, Joyce Poole, and Ian Douglas-Hamilton.

This work was greatly facilitated by the support of the Namibian Ministry of Environment and Tourism staff throughout the years, in particular: Werner Kilian, Pierre duPree, Johannes Kapner, Immanuel Kapofi, Rehabeam Erckie, Wilferd Versfeld.

I would like to acknowledge my first agent on this project, John Michel, who never left me in spirit but only left the industry, as well as Ann Downer-Hazell and Andrew Paulson, who had both turned from editor to agent at an opportune time. Acknowledgment is also due to Ken Wright for his encouragement during the brief window that he was my agent before returning to the editorial side of the table. The publishing world is indeed in flux. Thanks are owed as well to my extremely supportive editor, Christie Henry, at the University of Chicago Press, who saw many versions of this manuscript before it was ready for print, as well as my copyeditor, Yvonne Zipter, and the design team at UCP.

I also wish to express my gratitude to my steely penned companion, editor, etymologist, and father, Dan O'Connell, who helped me to stay on target with my deadline at all hours of the day and night; my mom, Aline, for her enthusiastic sharing of *her* steely pen with me; and my aunt Dorry, former New Jersey

Supreme Court Judge, for her vigilance and pursuit of clarity as a lay reader of the final versions of the chapters in this book. I also thank my brother, Dan O'Connell, CEO of HNu Photonics, for his support of my elephant dalliances, for his creative inspiration, and for collaborating on our vibrotactile hearing device and poaching detection device.

Lastly, I would like to thank Donna, the elephant, for the constant up close and personal reminder that elephants and humans have so much in common, along with the elephants of Mushara and most especially Greg for whom this book is dedicated.

This ongoing research was made possible by the support of Utopia Scientific contributing volunteer program participants, interns, and students (www.utopia-scientific.org). The Martha Daniel Newell Visiting Distinguished Scholar Program at Georgia College allowed me the time to finish writing this manuscript. Stanford University Vice Provost for Undergraduate Education (VPUE) faculty and undergraduate student grants from 2005 to 2014 were essential to the success of this research program. A generous grant from the Seaver Institute allowed me focused research time on this work early on, which helped build the foundation for a consistent research program, as did smaller grants from the National Geographic Society, Scheide Fund, and U.S. Fish and Wildlife Service African Elephant Research Fund. Support from the Oakland Zoo Conservation Fund and the Ndovo Foundation have also been important to our ongoing program.

Hormone analysis was done in Sam Wasser's Lab at the University of Washington and at the San Diego Zoo Institute for Conservation Research.

Much of this book had its genesis in articles and blogs that I have written, including:

"Casting Words in Nature's Best Light," *Writer Magazine*, November 2009

"How Male Elephants Bond," *Smithsonian Magazine*, November 2010, http://www.smithsonianmag.com/science-nature/how-male-elephants-bond-64316480/?no-ist

"Return to the Elephant Club," *Scientist at Work* (blog), *New York Times*, July 20, 2011, http://scientistatwork.blogs.nytimes.com/2011/07/20/return-to-the-elephant-club/

"Ritualized Bonding in Male Elephants," *Scientist at Work* (blog), *New York Times*, July 21, 2011, http://scientistatwork.blogs.nytimes.com/2011/07/21/ritualized-bonding-in-male-elephants/

"Carrots and Sticks in Elephant Land," *Scientist at Work* (blog), *New York Times*, July 25, 2011, http://scientistatwork.blogs.nytimes.com/2011/07/25/carrots-and-sticks-in-elephant-land/?_php=true&_type=blogs&_r=0

"Rules of Engagement in the Elephant World," *Scientist at Work* (blog), *New York*

Times, September 8, 2011, http://scientistatwork.blogs.nytimes.com/2011 /09/08/rules-of-engagement-in-the-elephant-world/?_php=true&_ type=blogs&_r=0

"Bromance and Bullies," *Africa Geographic*, March 2012

"Ranks in the Elephant Society," *Scientist at Work* (blog), *New York Times*, July 13, 2012, http://scientistatwork.blogs.nytimes.com/2012/07/13/ranks-in-the -elephant-society/#more-20418

"The Darker Side of Elephant Country," *Scientist at Work* (blog), *New York Times*, July 27, 2012, http://scientistatwork.blogs.nytimes.com/2012/07/27/the -darker-side-of-elephant-country/#more-20761

"Winding Down the Elephant Season," *Scientist at Work* (blog), *New York Times*, August 21, 2012, http://scientistatwork.blogs.nytimes.com/2012/08/21 /winding-down-the-elephant-season/

"The Meanest Girls at the Watering Hole," *Smithsonian Magazine*, March 2013, http://www.smithsonianmag.com/science-nature/the-meanest-girls-at -the-watering-hole-23122756/

"Family Strife," *A Voice for Elephants* (blog) *National Geographic*, July 27, 2014, http://newswatch.nationalgeographic.com/2014/07/27/family-strife/

"On The Nature of Family" (tentative title), *Slate*, forthcoming, spring 2015

CAPTIONS FOR
CHAPTER-OPENING PHOTOS

—————— ▼▲▼▲▼▲▼▲▼ ——————

"Kissing of the Ring": While Abe drinks, Willie places his trunk in Abe's mouth—
akin to a handshake, salute, or kissing the ring

"Journey to Mushara": Mushara tower and temporary field camp at Mushara water hole, a water point in the northeast corner of Etosha National Park that is
closed to tourism. The three-story tower provides a platform for tents, a research level, and a filming/recording level

"The Head That Wears the Crown": An extended bond group of elephants arrives
at the water hole on a hot afternoon before the full moon when the elephant
traffic picks up.

"Introduction to the Boys' Club": The young Congo greets the older Tim (*facing*)
and Willie. Some older bulls are more inclined to interact with younger bulls
than to ignore them. Congo would have recently left family in pursuit of com-
panionship among the adult male community, and not all adults are receptive
to taking youngsters in.

"Dung Diaries": I collect a fecal sample from Mike to assess his testosterone
and cortisol levels as well as to investigate his genetic relatedness within the
broader male resident population.

"Teenage Wasteland": Two young bulls test their strength against each other in a
sparring match, akin to an arm wrestle.

"Coalitions and a Fall from Grace": Second-ranking Mike (*rear view*) is met with a
coalition of resistance orchestrated by the third-ranking Kevin.

"Male Bonding": Bonded male associates often stand in very close proximity while
drinking.

"The Domino Effect": Greg, followed by Torn Trunk and Johannes, approach the
water hole where Smokey is keeping a close watch on their movements.

"*Capo di Tutti Capi*": Smokey, the most dramatic resident musth bull of Mushara,
is a very impressive figure. Unexpectedly, however, he is neither the most ag-

gressive musth bull nor does he shy away from social encounters with either younger bulls or other musth bulls, including Kevin. There are only two resident males that appear to irritate him, the top-ranking Greg and the half-sized musth bull, Ozzie.

"Of Musth and Other Demons": A showdown between two bulls in musth, Beckham and Prince Charles.

"The Emotional Elephant": Jack (*right*) engages in a trunk twist with Luke Skywalker (*missing right tusk*)—a vulnerable act of seeming affection only seen between the closest of associates.

"The Don Back in the Driver's Seat": Greg (*middle*), the don, flanked by his closest associates, who we have deemed the "boys' club"

"Closure": Mushara camp on a new moon night.

"Sniffing Out Your Relatives": Elephants have a keen sense of smell and often rely on scent and sound at a distance before using their vision to focus on something potentially dangerous.

"Where Are the Boys in Gray?": In very wet years, elephant ranges expand as there are many more places to drink, allowing them to search for food more broadly, and thus the return to Mushara is delayed until the ephemeral pans dry up.

"A Case for Dishonest Signaling": Musth bulls exhibit a suite of exaggerated behaviors that broadcasts their scent and that signals their high testosterone state.

"The Don under Fire": A surprise upset of the top-ranking Greg (*left, with head and shoulders low in submission*) by the occasional visitor, Marlon Brando (*right, head up and ears out aggressively*).

"Black Mamba in Camp": Tim and I removing a cape cobra from the bunker with a handmade noose threaded through a two-meter PVC pipe.

"Baying at a Testosterone-Filled Moon": Two coming-of-age bulls at full moon.

"Relentless Wind" Greg approaches the water hole at the end of the season when the wind and the quelea birds are at their peak.

"A Deposed Don": Greg soaks his trunk wound in the pan for long periods, while Keith often waits for him at the edge of the clearing.

"The Don Returns": Greg returns to Mushara at the beginning of the 2011 season, having regained his fitness and his rank, although the hole at the bottom of his trunk will never close.

"Scramble for Power": In the absence of the don of the boys' club, Prince Charles softens his bullying ways and climbs in rank.

"The Royal Family": She-Ra and her contingent of the Warrior family approach the water hole for a drink on a hot afternoon.

"Wee Hours": A twilight encounter with Congo.

"The Politics of Family" As a large group of elephants nears the water hole, they separate out into discretely ranked families, Wynona's family (*far left*) heading to the lowest-ranking drinking position at the pan, while the highest-ranking Slit Ear family (*far right*) heads to the freshest water position.

"A New Beginning": Wynona's new baby, Liza.